QUAKELAND

QUAKELAND

ON THE ROAD TO AMERICA'S NEXT
DEVASTATING EARTHQUAKE

KATHRYN MILES

DUTTON

DUTTON

An imprint of Penguin Random House LLC
375 Hudson Street
New York, New York 10014

LIBRARY OF CONGRESS CATALOGING-IN-PUBLICATION DATA
Names: Miles, Kathryn, 1974–
Title: Quakeland : on the road to America's next devastating earthquake / Kathryn Miles.
Description: New York, New York : Dutton, [2017] |
Includes bibliographical references and index. |
Identifiers: LCCN 2017011994 (print) | LCCN 2017022673 (ebook) |
ISBN 9780698411463 (epub) | ISBN 9780525955184 (hardcover)
Subjects: LCSH: Earthquake prediction—United States.
Classification: LCC QE538.8 (ebook) | LCC QE538.8 .M5387 2017 (print) |
DDC 551.220973—dc23
LC record available at https://lccn.loc.gov/2017011994

Printed in the United States of America
1   3   5   7   9   10   8   6   4   2

Set in Sabon LT Std
Designed by Cassandra Garruzzo

For Jane and Mike, my bedrock

CONTENTS

THREE

THE VOICE OF RAGE AND RUIN

It cannot be expected that an examination of the phenomena connected with this disturbance will throw any light upon the origin of earthquakes. Scientific men do not seem to have gained from the terrible earthquake that devastated the island of Ischia, one year ago, or from the great catastrophe of Krakatoa and the Straits of Sunda, one month later, much useful information in addition to that which they already possessed upon which to form new theories or reconstruct old ones. Whatever may be the cause—whether it is the falling of the roof of vast subterranean caves or the raging of a molten sea in the interior of the earth or the contraction of portions of the earth's crust caused by the gradual cooling of the entire ball—this fact is plain enough to unscientific persons, that such disturbances are beyond the control of men, and that the dangers that accompany them are best avoided by living beyond the limits of the earthquake and volcanic regions. Those who look back upon the experience of yesterday, either with fear or with curiosity, may well rejoice that their lots are cast in a country where earthquakes do no harm.

The New York Times, *11 August 1884*

PROLOGUE

Two years ago, a good friend returned from a mapmakers' conference with a gift for me: an enormous poster of the world. On it, the oceans are the deepest black a printer can print. The continents are just a shade lighter. What illuminates the map itself is a rendering of every earthquake that has occurred in the past decade: white dots for the really big ones, yellow and orange for the moderate ones, red for the rest. Together these dots create arcs of fire that define the edges of all the planet's tectonic plates. Plumes of light point to active zones in California and Alaska, Japan and India. You'd expect that, of course. But here's what that map also shows: burning embers up and down the East Coast of the United States and across Europe. Big patches of heat in the American heartland, along with Antarctica, Australia, and Greenland.

To my mind, this map best illustrates one very important fact: Earthquakes are everywhere.

Prior to writing this book, I had experienced two observable ones in my life, though neither of them really counts. The first occurred the summer I turned thirteen. My family was living in the

Quad Cities then, and I had acquired my first job, which was selling subscriptions to the daily newspaper. Each day, a van would collect me and a half dozen other adolescents. The driver would give us each two dollars for dinner at McDonald's, and then we'd be dropped off in a different neighborhood, where we were supposed to knock on every door and sing the praises of the paper to whoever answered. As it turned out, pretty much everyone who wanted to read the paper already had a subscription. So I took to pocketing my dinner money, ditching my bag of complimentary papers, and, instead of attempting sales, wandering up and down residential streets, imagining what life was like in cluttered duplexes and gated colonials. On one particularly sultry evening, I got back to the van just in time to hear that a strong earthquake had rattled the region: magnitude 5.1, the largest quake to hit the Midwest in over twenty years. Fans at a Detroit Tigers game had thought the stadium was going to collapse. Employees in a St. Louis high-rise reported that the building swayed back and forth. An entire apartment complex all the way over in South Carolina had been evacuated. I had felt absolutely nothing. In the days after, I replayed the evening over and over again in my mind, looking for some evidence that I'd been in a quake. Hadn't I felt a shake? Heard dishes rattle in a nearby apartment? Seen power lines ripple? In truth, I hadn't seen or heard anything—not even a dog's bark. I felt cheated.

The second quake occurred a few months before my thirtieth birthday. I was lying on a beach on Hawaii's Big Island, waiting out the appearance of a very stubborn great white shark nearby. Suddenly, the ground beneath me shifted an inch or so forward and then back. I looked up to see if anyone else had noticed. If they had, they didn't let on. Back in the hotel, the front desk clerk just shrugged and said, "Earthquake," when I asked about it. Apparently, when you live on a giant volcano in the tropics, tremors are about as noteworthy as fresh bananas and sunny days.

I guess it could be worse. A neighbor of mine experienced a

moderate quake in Maine during the throes of an ugly divorce. While he was working in the basement, he heard the dining room table screech across the floor and assumed it was his soon-to-be-ex-wife helping herself to the furniture.

Most Americans have earthquake experiences as lackluster as these. The country hasn't had a big one since 1989, when a violent quake rocked Northern California just as TV viewers were tuning in to watch Game 3 of that year's World Series. As luck would have it, the series was between the San Francisco Giants and the Oakland A's, which meant 67,000 people were in the stands instead of on highways and bridges. Hundreds of thousands more were at home, safely glued to their televisions. As it happens, every major network was also in the area, which meant that this quake—since named Loma Prieta—was the first truly televised one in history. Around the world, tens of millions of viewers (35 million in the United States alone) tuned in expecting pregame commentary. Instead, they heard the panicked shouting of announcers and watched as fans stumbled out of the stadium. They kept watching as helicopters circled a section of Interstate 880 that pancaked in the quake, killing over forty people and trapping dozens more. They looked on with disbelief at footage of the Bay Bridge, where a fifty-foot segment had broken off and, incomprehensibly, a twenty-three-year-old and her brother attempted to drive over it. The car hung off the edge of the bridge for what seemed an eternity as a coast guard helicopter rescue team risked their lives to save the passenger (the driver died at the scene).

Those images scared the bejeezus out of many of us: Here was proof of what an earthquake could do to our nation's infrastructure—even to the engineering marvels that are our modern highways and bridges. There have been other tragic images since then: photographs of rescuers pulling children out of unimaginable rubble in Haiti. Of climbers and Sherpas caught in avalanches on Everest. Of an eighty-foot wall of water destroying towns and resorts in Thailand. Of villages in Japan evacuated because of nuclear radiation. All caused by earthquakes.

So far, the twenty-first century has proven a particularly deadly one worldwide when it comes to ground shaking. The 2004 Sumatran earthquake and tsunami killed an estimated 228,000 people. A 2008 quake in China took 88,287 lives. In 2010, nearly a quarter million people died when a magnitude 7.0 quake struck Haiti. Factor in several other quakes with death tolls in the tens of thousands, and you have approximately 800,000 lives lost to earthquakes in this century alone. The only other natural disaster this century that comes even close to these numbers is Cyclone Nargis, which killed approximately 138,000 Myanmar residents in 2008.

Collectively, we have a poor memory when it comes to disasters like these. A few years go by; the images fade. As they do, we get lulled into complacency. We tell ourselves that quakes happen in other places—that they'll never happen to us. We've been lucky so far: In the United States, only about 150 people have been killed by earthquakes in the past fifty years. Nevertheless, even these moderate quakes have come with a huge price tag: over $30 billion by some estimates. And whether you've felt the tremors or not, there's no denying that every single state in the country has witnessed an earthquake—even seemingly unlikely places like Florida and North Dakota.

The US Geological Survey is the federal organization that oversees geological concerns—including earthquakes—in this country. According to the USGS, over 75 million Americans currently live in "areas of significant seismic risk." Already, those areas include thirty-nine states. Both of those numbers are expected to rise in coming years.

Meanwhile, on any given day, our planet experiences 1000 or so observable earthquakes. In 2016 alone, the US National Earthquake Information Center registered over 20,000 of them. Even the moon experiences major quakes (we know this because NASA once had a pretty robust system of seismometers up there measuring them).

Sometimes, these quakes are where we expect them to be, like California or Chile. Often they are not. Just ask the residents of Plainfield, Connecticut. The town hadn't seen any seismic activity since 1975, and even then it was so intermittent and minor it barely warranted mentioning in local geological annals. However, in the early months of 2015, Plainfield experienced over 300 quakes. Experts still can't say why. Nor can they say why it took them until August the next year to discover a major fault in Virginia, the same state that, in 2011, witnessed the largest earthquake east of the Mississippi since 1897.

The USGS is persnickety about details, records, and definitions. When it comes to earthquakes themselves, they define the phenomenon as a "sudden slip on a fault, and the resulting ground shaking and radiated seismic energy caused by the slip, or by volcanic or magmatic activity, or other sudden stress changes in the earth." A fault is any fracture or crack in the earth's crust where movement can occur. The USGS has mapped 2100 faults where earthquakes have occurred in what is now the United States. These faults come in every possible flavor and variety, from tiny cracks deep underground to massive fissures that can be seen from space. Some, like the ones composing the San Andreas Fault Zone, have left legacies well-known to many of us. Other faults, like the thousands running under New York City, are real mysteries—even to the scientists who study them.

It's Art Frankel's job to map these potential hazards for the USGS. He says what worries him most of all isn't San Andreas or even the faults under our most populous city. Instead, it's the faults we don't yet know about—and he says there are far more of them than the 2100 we've recorded. Each of these undiscovered faults also has the ability to rupture—we just don't know if and when it will.

The fact of the matter is that, no matter where you're reading this book, there's a good chance you could throw a rock and hit a fault. What's even more disturbing is the recent revelation that we don't even really need active faults to produce earthquakes: As it turns out, we're pretty good at doing it ourselves. Oil extraction, dams and reservoirs, wastewater injection—even the construction of an apartment high-rise in Taiwan—have all produced quakes in recent years. These induced tremors are enough of a problem that the USGS now includes them in their hazard maps.

Even so, determining the risk still involves a lot of guesswork. As far as scientific fields go, seismology is very much a new one. It's also an immensely frustrating one, if for no other reason than that seismologists often can't see what they're supposed to be studying. In fact, if seismologists are certain of one thing, it's that earthquakes remain the least understood of all our natural disasters.

Think about it. Meteorologists have hours to plan reconnaissance flights into hurricanes, where they can gather firsthand information ranging from barometric pressure to wind speed. Volcanologists have figured out how to measure ground deformation and vapor releases to predict eruptions days and sometimes months in advance. Even tornado chasers get enough warning to race across the prairie in pursuit of their next big storm and the data it can offer. Once they find a storm, they'll have about ten minutes to watch that funnel cloud do its damage. Hurricanes rage for days. So do forest fires. The average earthquake lasts about thirty seconds. Even really big ones are done in under five minutes. The devastation they can leave in that time is unparalleled. And unlike disasters such as hurricanes and tornadoes, earthquakes don't appear to be restricted to a particular geography or climate.

They are also incredibly expensive.

As our population grows, so, too, does the financial and personal impact of a quake. The Federal Emergency Management Agency (FEMA) estimates that, today, a single large quake in a metropolitan area could result in damages of anywhere from $100 to $200 billion. That includes direct losses, like crumpled bridges and toppled buildings, and also indirect losses, like lost wages and revenue.

Want to put that in context? Consider this: Losses from 9/11, the nation's deadliest and costliest terrorist attack, added up to $120 billion. There's a good chance our next quake will cost more.

What makes it so difficult to plan for that quake is the fact that we have no idea where and when it will occur. That fact also makes earthquakes painfully difficult to understand, especially since the scientists who study them are rarely anywhere in the vicinity of an earthquake when it occurs. And they are the first to admit: There's just so much we don't yet know about this phenomenon. Every seismologist has his or her own analogy to express this: "We know more about _____ than we do about the inner workings of our own planet." They fill in that blank with things like "distant nebulas" and "the Higgs boson particle" and "what hat Kate Middleton will be wearing to the races."

That's a particularly big problem in a country like ours, which is to say a country defined in so many ways by not only a complex infrastructure, but also an aging (and often decrepit) one. The United States is a huge place—3.797 million square miles, to be exact. We rely upon over 300 ports to receive goods from around the world. Those imported products, along with our domestically produced ones, get shipped on hundreds of thousands of miles of roads (the US interstate system alone comprises over 46,000 miles). They travel across bridges, under overpasses, and through tunnels. We also rely on our airports not just for our own travel but for much of our business as well. Our single biggest freight mover—FedEx—moves about 12 million packages a day, and many of them spend the night at FedEx's Memphis hub, which also happens to be the site of one of our nation's largest and most damaging fault systems. And then there's our water and electricity, our nuclear power plants and oil trains: They've frankly all seen better days.

Every four years, the American Society of Civil Engineers produces a report card evaluating America's infrastructure. For two decades now, the United States has maintained a solidly D average.

In 2013, inland waterways and levees got a D-minus. Dams and aviation and roads and wastewater systems all squeaked by with a D. Our highest grade? A B-minus for solid waste. Definitely not Ivy League material.

This is a big problem for all kinds of reasons, not the least of which is that it makes it less likely that any of these systems will withstand a major seismic event.

I wanted to understand how this could be. How we could know so little about our planet and the risks it poses to all of us? So I set out on a road trip across America to see just how unstable we really are. Over the course of a year, I drove around the country and visited casinos, temples, and geothermal plants. I stood outside nuclear power plants built directly upon faults and toured buildings no engineer thinks will survive a quake. I descended 7500 feet into silver mines, crawled through Hoover Dam's turbines, and hopped aboard steamboats traveling down the Mississippi. I dug up sediment samples with paleogeologists in Seattle and crushed rock in New York City. Along the way, I learned about earthquake triggers, forecasting, and damage rings.

Quakeland is an account of that trip. It's a story about the earthquakes that have already defined America and the ones that are still coming. Its cast of characters are the scientists and emergency managers out there trying to keep us safe—the ones plumbing the depths of glaciers in Antarctica and wrapping bridge embankments in Memphis and petitioning your elected officials to let them set off a couple of quakes so they can study how they work. They are fascinating people. And the work they're doing may just save your life.

ONE

LAND MADE
FOR YOU AND ME

THEIR CAMPSITE, OUR CORE

Picture a campsite—one of the standard-issue kinds replicated in national forests and parks around the country. In a lot of ways, it's the quintessential American image: the kind of image you'd find on a postage stamp or in an ad for a new made-in-the-USA truck or maybe even in your favorite *Brady Bunch* episode. In so many ways, these sites represent our collective values: After all, these campgrounds exist because of good national policy, good government—real bedrock stuff.

At the particular site I have in mind, there's a little gravel pull-off for a vehicle and a bare pad just large enough for a four-person tent. Off to the side, a picnic table cozies up to a well-used fire ring lined with blackened stones. Maybe you've slept at sites like this. Perhaps you and your kids have roasted marshmallows in that fire ring, feeling the chipped paint of the picnic table on your bare legs as you stretched toasting sticks into the embers, liking the way the soles of your shoes heated up on the rocks.

Even if you haven't actually done this, I bet you can still imagine it. Why? Because it is a deeply American scene: "Go West, young man" meets wholesome family togetherness, complete with hot dogs and Jiffy Pop.

Now imagine the force required to cleave that land—to strike a fissure between that picnic table and fire ring. One moment, the two objects sit side by side, just a few feet separating them. The next moment, the fire ring and everything to the east of it drops twenty feet. Your toes are no longer touching warm stones. They're dangling off a cliff.

This scenario sounds like the sort of thing that can only happen in Hollywood. But sixty years ago, that's precisely what happened at the Cabin Creek Campground, just outside Yellowstone National Park. On August 17, 1959, a magnitude 7.5 earthquake ruptured with an epicenter just a few miles from there. The sheer force of that rupture caused the land to split and half of the site to plummet in what geologists call a scarp, or a sharp line of cliffs caused by seismic activity. As terrifying as that would have been for anyone there, it was the least of the damage to the area: Just across the river, the nation's biggest recorded rock slide—over 73 million metric tons of debris—roared down a canyon wall, burying nineteen people. They were never recovered. At least nine other victims died of their injuries. Today, a massive lake rests where their campsites once stood. Nearby, you can find geysers and fissures and sinkholes where there were none—all because of this single earthquake.

You can be forgiven for thinking that the ground beneath your feet is solid. For most of us, it certainly appears as much. When I was seven, I ran away with my mother's formal silverware, wrongly thinking there was a utensil shortage in China. Because I knew that the shortest distance between two points was a straight line, I also took my dad's shovel, figuring I could dig my way there in a couple of days. I lasted about an hour before I hit the dense clay that makes up so much of the Mississippi River corridor. To my scrawny arms, it was impenetrable. I assumed everything below it was equally un-yielding. And so I returned home, defeated, before anyone had even noticed that the silver and I were missing.

If basic earth science had been covered in my first grade curriculum, I'd like to think I wouldn't have bothered with the shovel. Maybe you remember that iconic drawing of the planet cut into a cross section that's often used in geology textbooks. I always think it looks like a peach: thin outer skin, flesh that's not quite solid or liquid, a tidy pit. That outer layer, insofar as the planet is concerned, is called the crust. This is a misleading term. It's actually made up of a bunch of jagged pieces known as plates. The current best guess is that, right now, there are fifteen big ones and a handful of smaller ones floating around. The thicker and less dense ones are known as continental plates. The heavier ones are oceanic. We'll talk more about each in the next chapter.

What's important to know now is that these plates, which are forever in motion themselves, also bear the scars of billions of years of upheaval and trauma on this planet. For instance, beginning near Lake Superior, there's a 1000-mile forked crack down the center of this country known as the Midcontinent Rift. One tine of this crack snakes down to Oklahoma. The other works its way clear to Alabama. This rift exists because, about a billion years ago, the continent began to break apart. Scientists aren't sure why the rift began to form, though they think it might have been because of volcanic activity below it. Even more puzzling to them is why we're still in one piece (a similar rift threading through East Africa appears to be succeeding in ripping that continent in two).

Other marks are no less monumental. The Appalachian Mountains, the oldest on the planet, are actually scabs from a head-on collision of two plates. They are composed largely of rock that once made up the seafloor—and they were once taller than the Himalayas. This kind of dramatic shifting happens on our planet all the time. While I was writing this chapter, a new island, about a half mile wide, popped up just south of Fiji, thanks to an active volcano there. Another one appeared off the coast of Japan in 2013. Meanwhile, a new plate appears to be forming below the Indian Ocean, perhaps birthed by the 2012 Sumatran earthquakes.

While all this activity is visible on the crust, most of it is actually caused by what lies just underneath: the mantle, which is a combination of solid and liquid rock. The mantle makes up most of our planet's volume, and there's a huge variation in its temperature from top to bottom. Up near the crust, it's a cool 1800°F. As you plunge deeper, it reaches a temperature of almost 7000°F. (Silver, incidentally melts at 1763°F, which is just one reason why my chosen path to China was a bad one—at least where the longevity of my mother's salad forks was concerned.)

This difference in temperature between the outer and inner mantle creates a dynamic heat exchange as hotter rock and magma rise to the surface and cooler rock falls back down. Want to see this in practice? Think of a lava lamp. Or, if that's too groovy for you, dump a can of minestrone into a pot and watch it boil on the stove. There's a certain rhythm to the rotation of carrots and macaroni as they are pushed to the surface and then back down again to the bottom of the pot. It's mesmerizing—at least until you remember that we're floating on top of a very similar process.

Soup eaters or not, seismologists love the mantle. It's where everything happens in one big, dynamic mess. Parts of the mantle are solid. Some of it is plastic or even viscous. Its movement is responsible for the drifting and colliding of our plates. Earthquakes occur there.

Below the mantle lies the core, which is probably the least interesting layer for any book about seismicity. First discovered in 1936, it's also the least understood. What is known is that the core is made up mostly of nickel and iron and is divided into two parts— the outer, which is molten, and the inner, which is solid but only because it is under immense pressure. It is also the very hottest part of our planet—scientists think its temperature is probably between

9000 and 13,000°F (by comparison, the sun's chromosphere—the deepest layer we can currently observe—ranges between 6700 and 11,000°F).

To reach the middle of the earth's core, you'd have to travel down about 3400 miles. Then you'd have to go back up about another 3400 miles if you wanted to pop out on the other side of the planet. (It's about 6500 miles from Davenport, Iowa, to Beijing overland, which is yet one more reason my silverware reallocation plan was a lousy one.)

If you've been counting, I've already used the word "about" eight times in this chapter alone. It's a word you hear even more frequently among scientists who study the inner workings of our planet. In the past decade or so, they've come up with some sophisticated tools to help them visualize what's going on below us, including 3-D imaging techniques. Nevertheless, saying this instrumentation gives us any kind of definitive knowledge about what goes on inside the earth is a lot like saying you've mastered the inner workings of a human body because you've seen an X-ray or a CAT scan. You might have a decent idea of what's in there, but you're missing a lot of nuance.

Where geology is concerned, these gaps are not for lack of trying.

In 1958, a group of American scientists attempted to drill into the place where the crust and mantle meet, which also, incidentally, is the place where most of our seismic activity occurs. Geologists call this boundary the Mohorovicic discontinuity, so named for Andrija Mohorovicic, the Croatian seismologist who discovered it in the early 1900s. (Pro tip: If you want to look like you know what you're doing at a seismology conference, refer to this space as "the Moho.") Like the crust, the Moho is far from uniform. On average, it tends to sit about five miles below the ocean floor and twenty miles below any continent. If you want to find it, you'd be much better off looking under an ocean instead of, say, Cleveland.

At least, that's what Walter Munk figured. A Vienna-born scientist who studied in the United States and applied for American citizenship after the Anschluss, Munk made a name for himself in World War II by correctly projecting wave heights, which allowed more than 160,000 Allies to land on the shores of Normandy during D-day. In the years immediately after the war, Munk played an integral role in our nation's early nuclear testing on Pacific atolls. His job? To look for and measure tsunamis created by the blasts. After that gig, Munk proposed Project Mohole: an attempt to drill into the Moho layer at depth. The project was endorsed by the American Miscellaneous Society, the loosest possible consortium of scientists: AMSOC, writes geologist and oceanographer Kenneth Hsu, "had no statutes or bylaws, no official membership, no officers, no formal meetings, no proceedings." Any scientist could become a member merely by saying so. The primary function of AMSOC was to support projects that had been rejected by other agencies, most notably the Office of Naval Research. Munk figured they were just the group to get behind his project. And so, one morning in the spring of 1957, he invited a few geology and oceanography friends over for breakfast. They appointed themselves an official AMSOC subcommittee. Together, they also submitted their Mohole proposal to the organization as a whole. The rest of AMSOC loved it. Theirs was maybe not the most prestigious endorsement for a project, but it nevertheless helped persuade the National Science Foundation to fund Project Mohole.

As presented, Munk's plan was a three-tiered attempt to bore below the earth's crust off the coast of Guadalupe, Mexico. This proposal raised no small amount of controversy in the geological world. Young radicals like Hsu tended to love it. Members of the establishment thought it was an absurd waste of money.

The Mohole group managed to overcome enough dissent and technical difficulties to begin their project. They succeeded in drilling down through the ocean floor about 600 feet before internal conflicts in the Miscellaneous Society led to its disbandment. The National Academy of Sciences took over the project in 1964. By then, the project's tab was over $1.8 million. Estimates for what it would take to finish the job were $68 million. That was a hard number for Congress to swallow. They cut off funding two years later.

The Soviet Union began a similar project in 1970 known as the Kola Superdeep Borehole. They made it down about 7.5 miles before incredible heat (about 350°F) basically melted their drill bits. Team scientists tried a variety of different solutions, but eventually the problem of temperature proved too much.

There is no easy way to get to the Kola Peninsula, which lies near the top of Finland and entirely above the Arctic Circle. If you're willing to take eighteen hours' worth of flights on increasingly dubious-looking airplanes, followed by a very confused SUV ride across the tundra, you'll eventually reach the weirdly bombed-out remains of the dig site, which includes no small amount of discarded equipment and trash littered around. The actual borehole, which is just nine inches wide, is covered by a rusted cap. At the end of the project, engineers welded it shut and added twelve enormous bolts for extra measure. Even if you were able to remove them, you wouldn't find the Moho: The borehole got only about a third of the way down.

———

If you really want to see what's going on inside our planet, your best bet is to go to places like Yellowstone National Park. There, you don't have to drill down thirty miles to see the guts of the planet. Instead, they come to you. The park itself actually sits in a caldera, a giant sinkhole created during a massive volcanic eruption. The caldera is what scientists sometimes call a "supervolcano." While not actually a scientific term, that is the word scientists use to categorize any volcano capable of spewing a trillion tons or more of ash and debris in an eruption. Worldwide, there are only about a dozen supervolcanoes capable of that kind of explosion. Yellowstone is one of the biggest. Just over 2 million years ago, it erupted and spewed more than 585 cubic miles of magma—about 200 times more than Krakatau in its epic 1883 eruption. That same Yellowstone eruption left an enormous ash bed that blanketed much of what is currently the United States. Think of an equilateral triangle with its top in Winnipeg and two bottom corners in San Diego and New Orleans, respectively. That's how far the ash spread.

Fueling Yellowstone today are two reservoirs of magma. Geologists call these reservoirs "blobs" (that's a technical term). They're still not sure just how much magma is in those reservoirs, but current best guesses are that the two blobs probably total about 35,000 cubic miles of magma. For comparison's sake, it takes about 2900 cubic miles to fill Lake Superior.

In general, magma tends to stay deep within the earth. Most of our planet's volcanoes occur in places where two plates meet and one slides under the other (what geologists call a subduction zone). That sliding melts the otherwise solid mantle and allows a column of magma to escape. The Pacific Ocean's Ring of Fire is a great example of this process in action. So is Mount Saint Helens, which is also located on a subduction zone. Yellowstone, on the other hand, is smack-dab in the middle of a continent. So how did it come to be?

Jake Lowenstern is the scientist-in-charge of the Yellowstone Volcano Observatory, the organizing body that oversees the seismic network and other research going on there. After receiving his bachelor's degree, he spent a year in Sicily, home of Mount Etna— one of the world's most active volcanoes. In 1986, it erupted in spectacular fashion, pouring lava for 120 days in what one Italian volcanologist called "effusive activity." Lowenstern was captivated.

Before taking over at the YVO, Lowenstern dedicated his time to studying the chemistry of hydrothermal systems and their relationship to magma. It was a quiet job with a lot of time in the field, and mostly consisted of taking samples of odiferous gases that make most people want to race back to their cars. He sometimes misses those days. Lowenstern is a slight guy with a big head of hair and a crooked smile. Even when he wears a salt-and-pepper beard, he still looks like a college student. He has unlimited patience and goodwill, both of which are required of his new job—and in ways he couldn't previously imagine.

"Scientists have been interested in Yellowstone for nearly sixty years now," he says, "but it took us forty to get the general public's attention. Now that we have it, we're spending most of our time keeping things accurate—especially for the media."

Volcanoes like the one at Yellowstone are caused by hot spots: places where weaknesses in the mantle allow magma to burn through, creating a fountain of lava called a plume. Just like a faucet channels water from your pipes to your sink, a hot spot channels magma from deep into the earth up to the crust. Once there, that plume works like a blowtorch, burning away everything that crosses its path, including the underbelly of the earth's crust. As it does, it belches lava up and onto the plate. What results are land formations like the Hawaiian Islands. They're located in the middle of the Pacific Plate, which has been slowly creeping to the northwest for hundreds of millions of years. As that plate moves across the Hawaiian hot spot, the plume burns away the deep oceanic crust and sends up a spray of lava, which quickly hardens. You can trace the movement of the plate over this spout by radiometric dating each of the resulting islands. Kauai, which sits at the top of the chain, is about 5.5 million years old. By comparison, the Big Island of Hawaii, which is the most southeastern island, is a relative infant: Its oldest rock is just 700,000 years old, and the island is still very much being built, thanks to Kilauea, the active volcano that currently defines (and is building) the island. Fast-forward another million years or so, and the state may well have an additional island to its south.

The hot spot currently below Yellowstone functions in very much the same way. The park sits at the top of an arc that runs northeast from the border of Nevada and Oregon. Surrounding that arc are the dramatic mountains that comprise the Basin and Range Province, a huge swath of land that descends from Oregon and Idaho down into Mexico. Between them lies the Snake River Plain, a strangely level strip of rock. Mountains used to stand there. Each one was burned away by the Yellowstone hot spot, which left in its wake a series of volcanoes that either eroded or were blown apart and, in either case, were eventually covered with a sheet of smooth lava. Twin Falls, Idaho, sits atop the remnants of one such volcano that was active about 6 million years ago. The area around Jackson, Wyoming, was atop that same plume 2 million years ago. If tectonic movement continues in its current direction, the area currently known as Billings, Montana, will be obliterated by the hot spot in a couple of million years. Eventually, the hot spot will sear into Saskatchewan, too.

Right now, the magma pushed up by the hot spot continues to burn and shape Yellowstone National Park. In some places, it smolders just three miles below the surface. It's this magma that fuels the geysers and hot springs and mudpots that bring millions of visitors to the park each year. It's also responsible for the thousands of earthquakes occurring there annually.

Based on dating of the rock in Yellowstone, scientists like Lowenstern estimate that, since that last big eruption at Yellowstone 2 million years ago, the volcano has relieved pressure with smaller lava flows, thought to occur every 20,000 years or so. The last one scientists can document happened more on the order of 70,000 years ago. In other words, the Yellowstone Caldera is currently in what volcanologists call an "eruptive lull." And this is where Lowenstern's job starts to require even more patience and goodwill.

"The big question, of course, is whether the volcano is going to start up again and, if so, when it'll start up again," he says. "Will it be a big eruption, or has Yellowstone stopped altogether, or is the hot spot going to start up again somewhere else?"

Most likely, he says, there will be another lava flow like the last one, which oozed over the southern base of the caldera. If that happens in our lifetime, it'll undoubtedly make a mess for the park: Roads will be covered and rendered impassable, forest fires will break out, animals will lose key habitat.

But that kind of scenario lacks the punch TV writers and producers often look for. They want something bigger—something headline-producing and worthy of a film starring Bruce Willis or Vin Diesel. They were certain they had it in 2003, when a flurry of geological activity, including earthquake swarms and the reactivation of some dormant geysers, made it seem to the popular media like the big one could be coming any second. Internet forums went crazy with doomsday scenarios and conspiracy theories. Even the normally staid BBC and National Geographic produced sensationalized documentaries predicting massive eruptions. Lowenstern went into damage control mode, fielding countless interviews and trying to bring the story back to science, which exists in the realm of probabilities and statistical analysis: *Probably* there's not going to be a major eruption at Yellowstone in our lifetime. *Probably* there won't even be a catastrophic lava flow—*probably,* instead, just some forest fires and wrecked roads. That doesn't make for great storytelling, but it's enough for Lowenstern, who says he thinks of himself as a rational fatalist.

"Really bad things happen on a geological timescale," says Lowenstern. "There's not much we can do about that."

He says he's okay with the uncertainty that poses.

"Look," he says, sitting in his office in Menlo Park, California. "Imagine you're a mayfly born on a river in Colorado. You're around for a day. You mate and you die. Do you believe that there's such a thing as a snowstorm? Or skyscrapers? That's stuff you're never going to experience. Humans are like that with respect to earth processes.

As a species, we woke up five thousand years ago. We didn't start really taking notes until a few hundred years ago or so. We're trying to figure out things that are ten thousand times older than our collective consciousness. We're just opening our eyes and looking around, and yet we expect to have all the answers."

If Lowenstern is certain of one thing, it's that we're not even close to finding those answers. What we do have, he says, is the ability to estimate the likelihood of events coming to fruition. To that end, he likes to evoke "a certain former secretary of defense" (Lowenstern is always careful not to use Rumsfeld's name): There are known knowns and known unknowns, he says. Those are the easy ones. It's the unknown knowns and unknown unknowns that make his work challenging—particularly where public outreach is concerned.

"People don't think things are possible if they haven't experienced them," says Lowenstern. "And so they don't prepare for them, either. Geologists know it can happen, but we have a hard time persuading the general public, especially when we can't say whether or not it'll happen in our lifetime."

For Lowenstern, it all comes down to playing the odds. His lab is just miles from the biggest fault zone in the country. Unlike with Yellowstone, there's a really good chance it could have a major rupture in his lifetime. Still, he doesn't have earthquake insurance. "The premium is just too much," he says. "And even the best deductible is way too high. I did the math: I'm always going to come up short."

A BEAUTIFUL PLACE FOR
AN EARTHQUAKE

There's a famous saying often attributed to Mark Twain that goes, "If you don't like the weather in New England now, just wait a few minutes." Yellowstone has its own riff on this idea: If you don't like the landscape, wait a year or so. Thanks to the seismic activity there, the topography is always changing as different sections rise and fall to their own geologic tempo. That's true across the planet, of course. Here in the United States, the Mississippi delta sinks as much as ten millimeters each year, thanks to normal compaction of organic materials. Parts of New England are rising because of what scientists call "glacial rebound." Charting the seismic rising and falling of the Pacific Northwest is like playing whack-a-mole. Keeping track of Yellowstone is pretty much like playing the turbo version of that game—while you're drunk.

Bob Smith has been trying to do just that for over fifty years. Now in his late seventies, Smith spends most of his time at his ancestral home in Grand Teton National Park, where his forebearers homesteaded for generations. It's a quiet place, with limitless views of some of the most dramatic peaks in the Rocky Mountains. It's also an easy drive to the south entrance of Yellowstone Park.

Smith makes that drive a lot. And he has a special fondness for the roads leading into Yellowstone. In the early 1970s, he used them to establish just how seismically active the caldera really is. The first national park, Yellowstone was founded as a kind of sociological experiment: The region had no obvious agricultural or development value, and its remarkable geology was unlike anything explorers had seen. Congress approved its establishment as a national park in 1872 and appointed an unpaid superintendent to keep an eye on the place. No one expected just how popular it would become. By 1922—the park's fiftieth anniversary—approximately 100,000 tourists were visiting the park each year. Over half of them did so in their own cars. The long-worn roads of the rustic park were no match for this kind of traffic. And so Congress reluctantly passed a $7.9 million bill (the equivalent of over $100 million today) to modernize thoroughfares in this and other national parks.

Building a road on a caldera is no easy enterprise. To facilitate the project, the Department of the Interior dispatched a team of surveyors, who installed 300 benchmarks in the park, each bearing the coordinates and elevation of the spot where it lay. Over the years, Smith had seen many of them in his travels through the park. Like Lowenstern, he's a data guy: For Smith, there's a comforting certainty in statistical information. These benchmarks had that kind of detail in spades. There, emblazoned over and over again, was a window into the park's geological past. If, as geologists had begun speculating, the caldera was indeed capable of activity, he could use those benchmarks to prove just how much. Beginning in 1973, Smith and his students spent three summers resurveying the area. And what they found excited even the taciturn Smith: Since those benchmarks had been installed, the entire caldera had risen about a meter. In scientific terms, that's a really big deal. Smith still gets excited about it.

"I learned to fly an air force jet in 1963. A year later, I drove a dog team across Antarctica. Neither held a candle to that discovery," says Smith today.

What fascinated Smith even more was the discovery that this uplift was temporary. Beginning in 1985, Yellowstone witnessed a swarm of seismic activity—over 3000 earthquakes in a three-month period. That alone is pretty interesting. But what really got Smith's attention was what happened just after: The caldera reversed its direction and began to drop (geologists call this behavior subsidence). Over the next decade, the park fell about twenty-five centimeters. Then, inexplicably, it began to rise again. In 2009, the caldera went into overdrive and began rising as much as seven centimeters a year. That activity peaked in 2014. Currently, it appears to be subsiding again.

Smith doesn't know if and when this pattern will change. Most volcanoes tend to grow just before they erupt: Mount Saint Helens famously rose over 200 feet just before its 1980 eruption; in Iwo Jima, you can find enormous coral heads nearly 400 feet up in the hills—evidence of intense volcanic uplift there. The movement at Yellowstone isn't nearly as organized. Instead, portions of it rise and fall at will, earning it the designation "caldera at unrest."

The root cause of that unrest is the magma itself. As it moves around the caldera, the magma heats and cools, expanding and contracting. Meanwhile, the rocks closest to the surface remain brittle. Brittle objects don't take strain well. Consider what happens when a pebble dings your windshield on your way into the grocery store parking lot. You may see a nick when you grab your cart. By the time you've checked out, that nick will have become a couple of long cracks. The next morning, those cracks will have become a web you can't see around.

The planet is like that, too.

In geological terms, each of those windshield cracks represents a potential earthquake. And in Yellowstone, they spider with a frequency not often seen elsewhere. In 2010, another swarm resulted in over

2300 quakes. Even in a slow year, the park experiences about 2000 earthquakes annually, making it one of the most seismically active areas in the country. By comparison, all of Southern California—generally regarded as the most seismically active region in the Lower 48—gets about 10,000 a year. Alaska gets more than 12,000. If that doesn't seem stark enough of a comparison to you, consider this: Yellowstone is about 3400 square miles. Southern California is about 56,000 square miles. Alaska? Over 663,000 square miles.

Part of the thrill of earthquakes around Yellowstone is that they are so present. Quakes occurring on normal faults tend to be deep—sometimes as much as 300 miles below the surface. In Yellowstone, the brittle layer of rock goes down only a few miles before it becomes more ductile rock. That's close enough to the surface for us to hear the waves as they pulse through the ground.

Lewis and Clark had that experience while camping there in June of 1805. In their journals, the former waxed poetically about their dinners of buffalo and trout, the steep riverbanks and even steeper cliffs. Clark, the pragmatist, instead reported his men couldn't sleep and complained about a strange noise "as loud and resembling precisely the discharge of a piece of ordnance of 6 pounds at the distance of 5 or 6 miles." He initially wrote off the disturbance as thunder. The next day, while walking the plains, he heard the explosion again. There wasn't a cloud in the sky. He "halted and listened attentively about two hours." Somewhere in the distance, something boomed again. And again. Clark took out his pocket compass and tried to follow the noise. Was it water running through a cavern? No, he decided: "In such case the sounds would be periodical and regular, which is not the Case with this, being Sometimes heard once only and at other times Several discharges in quick Succession." After much searching, he eventually gave up. But as his team continued venturing westward, Native Americans he met along the way reported similar sounds throughout the Rockies. They, too, were unable to account for the noise.

Today, scientists believe these noises are indicative of a much larger seismic region extending from Las Vegas up to the Canadian border. To the east of this region sits the North American continental plate and some of the oldest, most stable rock anywhere on earth. To the west, chaos: the violent crashing of plates, the constant destruction and creation of new crust. Sandwiched in between is the Basin and Range.

Twenty million years ago, tectonic forces began to rip apart this section of the country, too. But unlike the area around Lake Superior, the Basin and Range merely stretched. As it did, the crust thinned. Magma—much like that in the reservoirs below Yellowstone—began to heat the region, cooking the crust and causing it to blister. Above, the cooling rocks grew brittle. Meanwhile, tectonic forces continued to stretch the region. Weaker areas in the crust cracked under pressure, forming faults. When too much pressure builds on either side of those faults, they rupture.

This is not news to residents of the Basin and Range, which is marked by hundreds of faults. The biggest—and most lethal—is the Wasatch, a 240-mile loaded fracture that runs from central Utah into Idaho. Salt Lake City was built directly on it. In 1883, G. K. Gilbert, a senior geologist with the newly formed US Geological Survey, wrote a stern letter to the editor of *The Salt Lake Daily Tribune*. He'd seen evidence of a giant fault—and what it could do—around the city. A major earthquake, he prophesized, was more than possible. "It is useless to ask when this disaster will occur. Our occupation of the country has been too brief for us to learn how fast the Wasatch grows; and indeed, it is only by such disasters that we can learn. By the time experience has taught us that, Salt Lake City will have been shaken down."

Since Gilbert's prophesy, geologists have found dozens more faults in the region. They, too, are capable of significant earthquakes. And Bob Smith has a hunch they are connected in ways we're only beginning to understand.

Twenty years ago, Smith published a geological history of the region that includes detailed instructions for a do-it-yourself seismic tour of the region. It's about to be reissued in even more detail. It's inconceivable to him that anyone wouldn't want to take this trip. He begins our first phone conversation by asking why we've never met in person. I have no good answer. Ten minutes in, he asks when I'm coming to Yellowstone. I tell him I'll hop a plane.

A week later, I learn another important fact about Yellowstone: It gets almost as much snow as it does earthquakes. Even my rented SUV is no match for it. And so what was supposed to begin as the classic American road trip instead begins with me playing stowaway in a 1956 Bombardier snowcoach. "Griz" runs on a combination of rear tracks and two sturdy skis up front. On the wet snow blanketing the park, the snowcoach sounds a whole lot like an industrial blender filled with Tonka trucks. But it does not waver. By the time we reach the Old Faithful Snow Lodge, we're in nothing short of a whiteout—so much so, in fact, that I can't make the short hike to the cabin where I'll be staying. And so, as the other passengers of the snowcoach continue their tour of the park, I instead hunker down near the lodge's iconic fireplace until it's safe to venture toward my cabin. Outside, even through the blowing snow, I can hear the hiss of Old Faithful and the dozens of other geysers around me. It is the sound of something undeniably alive.

———

Myrtle Painter had a far more dramatic reaction when she visited Yellowstone in August of 1959. The forty-two-year-old mother of four had come to the park with her husband, Ray, and their three daughters. Their eldest child, a son, couldn't get away from army duty to meet them. And so the five remaining Painters packed up their dog, Princess, and towed a new camper to the park. Myrtle was uneasy long before they arrived. Before departing their home in Ogden, Utah, she'd instructed the girls to clean the house from top to bottom; she also insisted that Ray make sure all of their insurance—including their life insurance—was in order. When she worried aloud about Princess making the trip and possibly getting lost, Ray commandeered an old dog collar left at a gas station. The attendant said that the previous wearer had been hit by a car, but the collar looked like new, so Ray snagged it anyway. It never occurred to him to change the tags.

Once in Yellowstone, Myrtle was inconsolable. She didn't like the boiling mudpots or the erupting geysers. Everything there was way too volatile. She wondered aloud how many earthquakes the place must get—it had to be a ton. After a couple of days, she finally persuaded Ray to pack up their camping gear. As they loaded up the car, she told her family they needed to hightail it out of there before the whole place blew up.

That was fine with Ray. An ardent angler, he was itching to get on the water. Rumor was that the cutthroat trout on the Madison River couldn't be beat. He drove the family, the overloaded car, and the camper out the park's west entrance and made a beeline for the river, not wanting to waste a moment. The weather that day couldn't have been better—cool and sunny, with the promise of a dramatic full moon later that night. The Painters skirted the town of West Yellowstone, then turned north on Route 191. In the backseat, the twelve-year-old twins, Anita and Anne, grew increasingly impatient. They wanted to visit the ranches and cafés dotting the highway; they'd never seen a huge earthen dam like the one that made Hebgen Lake. Over 700 feet long, with a concrete core nearly ninety feet high, it was an engineering marvel—especially at an altitude of 6500 feet. *Couldn't they stop?*

Ray said no. He wanted the perfect site and worried the campgrounds were already filling up for the night. And so the family kept driving deeper into the 300-foot canyon, by now following the meandering rim road as it flanked the widening river on one side and the sharp cliffs of the range on the other.

Eventually, the Painters arrived at Rock Creek, one of three campgrounds nestled between the highway and the mouth of the Madison River. Ray was right: The place *was* remarkably full for a Monday afternoon. The first site he chose was near the road. The girls unpacked the family's coolers and chairs and pots and pans; they let Princess sniff around. But Ray was restless. He worried the site would be too sunny for Princess to be left in the car, too far for them to walk with rods and bait boxes. He left to scope out better sites and eventually settled on one right near the river. Once he returned with the news, Myrtle and the girls packed up their gear, moved to the new site, and unpacked again. By then, the girls were tetchy. Myrtle cooked steak and potatoes over the fire, corralled the kids into bear-proofing the campsite, then tried to negotiate sleeping arrangements. Anita wanted to sleep in a hammock above her folks. Myrtle said no. A meltdown ensued. Eventually, both twins clamored into a bunk at one end of the camper. Anita gave Myrtle the silent treatment and refused to be tucked in—punishment for being denied use of the hammock. Myrtle and Ray took their usual spot at the other end. Carole, their eldest daughter, elected to sleep with Princess in the car. By the time all three girls were asleep, Myrtle was exhausted. She took a towel and bottle of shampoo and walked down to the river to wash her hair. No one heard her leave.

Ten miles northeast of the Painters' site, Route 191 intersects with Canyon Creek Road. On one side of the intersection rises the steep southwestern flank of Kirkwood Ridge. On the other lies scrubby pastureland. Ranchers in that area turn in early. So do anglers. Probably, then, by the time Myrtle left her camper, that intersection was as dark as the hills and trees surrounding it.

Underneath that quiet intersection, a fourteen-mile fault called Red Canyon had been straining for hundreds—maybe even thousands—of years, absorbing energy from the active Basin and Range.

At precisely 11:37 P.M., the Red Canyon Fault ruptured. It took with it the nearby Hebgen Fault as well.

What does this mean?

Remember: The earth's crust is in no way uniform. In some spots, like North Dakota, it's thick, deep granite—hard to move and even harder to break. In others, like Yellowstone, it's paper-thin. In most places, the ground is something in between. As stress builds up in those places, different rocks respond differently. Every type of rock has a breaking point—a place where it will fail or fracture. A fault is a site of such fracture where movement has occurred. Sometimes that movement is just a slow slip. More often, it's an abrupt and violent one. The rock lurches—often rapidly and with the force of hundreds or thousands of years of accumulated energy. Just how much of the rock moves—and how far it goes—will in large part determine the size of the earthquake.

The specific place in the rock where that rupture occurs is called the hypocenter. Often, the hypocenter is tens of miles underground. Draw a line from that hypocenter straight up, until you reach the surface of the earth. That's the epicenter.

So why do an earthquake's effects often extend for tens or even hundreds of miles in either direction?

Pluck a guitar string. Drop a heavy rock into a still pond. In both cases, the resulting disturbance is transported across space one particle at a time in an ever-widening series of waves or rings. An earthquake works the same way: Moving faults create a tremor, the energy of which radiates outward. Inside the rock, that energy moves as two interior, or "body," waves. The first is called a P wave (or compressional wave). A P wave moves in much the same way an acoustic wave does when you pluck the guitar string: It moves away from the object while shaking back and forth in the direction of that outward movement. The second seismic wave is called the S wave (or shear wave). It also emanates from the seismic body, but its shaking occurs perpendicular to that movement. P waves move faster than S waves. The longer the time between the arrivals of a P and an S wave, the farther away you are from the epicenter of a quake—not unlike what happens when we see lightning and then count the seconds until the subsequent thunderclap arrives. Sometimes, if the P wave is close to the surface, we can hear what sounds like an explosion as it disturbs the air in its path—just like when we hear a sonic boom as a jet flies overhead. The S wave is actually the much stronger of the two, and it makes plenty of noise as well. We just can't hear it because it's below the lowest range of human hearing. (S waves usually register around 5 hertz. The human ear bottoms out around 20 hertz.)

Just how intensely earthquake waves shake us depends upon a whole host of factors, including the type of soil they move through and the depth from which they emanated. The shallower the earthquake, the more damage it causes (one reason why the 2016 earthquake in Italy leveled so much of the area, even though it wasn't all that strong). The Hebgen quake's hypocenter was about six miles below the surface. Its first waves emanated from the canyon in ever widening circles. They reached Kona, Hawaii, and the Adirondack Mountains in New York less than ten minutes later. On the surface, two additional types of waves did their work as well. The ones called Love waves rocked the ground from side to side. The ones called Rayleigh waves rose and fell like those on a storm-churned ocean. Both reached the southwest corner of Yellowstone Park in seconds. There, over 500 guests and staff members were packed into a recreation center, watching as Jeannine Anderson edged out Louann Proffitt for the title of Miss Old Faithful.

Anderson had just begun her victory walk when the room began to quiver. Later, survivors would differ in their accounts of how long the shaking lasted: Some would say it was just a few seconds; others would swear it went on for minutes. A park bus driver said it felt like he had won free mamba lessons. Pageant participants and their audience were a lot less impressed. As the shaking continued, they tumbled out of the recreation center just in time to watch the inn's immense chimney crumble. It toppled into the dining room below, which somehow set off the lodge's entire sprinkler system. Hot-water pipes had burst as well, and were sending steaming torrents down the walls of guest rooms. Panicked, guests on upper floors jumped from their windows, some wearing only towels. They began to congregate around bonfires and in their cars, too afraid of aftershocks to return to their rooms. Around them, long-dormant geysers erupted and new ones exploded where, only hours earlier, tired families had stood.

All of that was a cakewalk compared to what was unfolding back on the Madison River. There, dozens of scarps formed as the ground collapsed. Our campsite—the one with the peeling picnic table and cozy fire ring—was one such place. Meanwhile, tremors exploded with the sound of more artillery than Lewis and Clark could have imagined. The Hebgen Dam's superintendent heard the blasts and assumed Russia had declared war. Around him, wooden homes were tossed off their foundations. Those made of stone collapsed. Pipes burst and fences snapped. Up and down West Yellowstone's main street, storefront windows exploded and shelves overturned. Boulders rained down like hail, obstructing roads and pockmarking fields. Sand spouts erupted hundreds of feet into the air.

That same energy dropped the basin of the lake twenty-two feet in places, sending water rushing from its banks and flooding trailers and cabins. The surge pushed debris high into the sagebrush surrounding the campgrounds. It topped the Hebgen Dam by at least three feet, then receded, then overtopped it again—four times in all. With each inundation, the caretaker swore the dam, now pushed askew by nearly three feet in places and bearing cracks six inches wide, would give.

Two miles downstream, witnesses watched as the lake rose twenty feet in fewer than ten minutes. Four miles downstream from them, Myrtle Painter still knelt at the bank of the river, washing her hair. The quake's surface waves knocked her over. A few hundred feet away, it rattled the camper and woke her family. There was a brief moment of stillness. And then the sound of a crash—louder than anything any of them had ever heard before. One resident said it sounded "like a thousand winds going through the trees." A camper called it the sound of the end of the world. On the south bank of the Madison, almost directly across from the Rock Creek Campground, the front face of one of the canyon walls calved from the stress of the earthquake. In eight seconds flat, 80 million tons of rock and trees (over 300 Empire State Buildings' worth) plummeted down the canyon, dammed the river, and buried much of the campground. The sheer force of the slide created a vacuum of air that sucked the clothes off Myrtle Painter. As the wind raced back into the canyon, it did so at hurricane strength. The Bennetts, a family of four, lay not far from the Painters: The night was so temperate, the kids had persuaded their parents to let them sleep in bedrolls outside their camper. They slumbered through the initial shaking, but their parents didn't. By the time they made it out of their camper to check on the kids, an enormous blast of air whipped through the campground. Mr. Bennett grabbed hold of a thin tree and was pulled as taut as an American flag in a stiff breeze before he lost his grip and was blown into the darkness. That same wind overturned cars. It drove twigs into the knee socket of another survivor.

The slide itself buried cabins and campsites; its errant boulders crushed tents and cars and the occupants inside. They came to rest just a few feet away from the Painter camper, where the twins huddled in terror. An instant later, an enormous tidal wave created by the falling slide surged through the campground, tearing the Painter camper in half. It swept Ray away from the trailer, knocking him against branches and splintered hunks of cabins, eventually pinning him beneath an enormous pine tree. That same wave engulfed Myrtle, slamming her against rocks and cars and other debris before depositing her on an enormous boulder. Her eldest daughter, Carole, who had weathered the quake, the slide, and the tidal wave un-harmed, staggered out of the car, looking for the rest of the family. She found her mother on that boulder. Myrtle had a collapsed lung and a compound fracture in her shoulder. Her arm was all but sev-ered below the elbow. She was losing blood. Fast.

Somewhere out in the darkness, Ray was faring only slightly bet-ter. When a group of survivors managed to extricate him, they found deep gashes down both of his femurs and at least one severed artery.

Ironically, the occupants of the Painters' original campsite were complete unharmed by the catastrophe. Carole, attracted by the headlights of their station wagon, helped lead Myrtle toward those lights. A makeshift field hospital was emerging there, with jeeps and tents serving as hospital rooms. A nurse who happened to be on vacation at the campground was busy making tourniquets and ban-dages out of flannel shirts and whatever else she could find. One of the survivors assisting her was a World War II veteran. Later, he reported the injuries he saw there were worse than any he'd seen during the war.

Fifty miles downstream, word reached the small town of Ennis that the Hebgen Dam had been so damaged by the quake it could break at any moment. Amid the wail of the lone emergency response siren, groggy families stumbled toward their cars with whatever they could carry and made for higher ground. By dawn, the town was deserted. Meanwhile, the dam created by the slide—over a mile long and 300 feet high in places—had impounded the river. Within hours, water levels there rose twenty feet. Within days, that newly formed lake would be over a hundred feet deep.

Residents in other nearby towns knew about the quake. They'd felt the shaking. Many had heard the enormity of the collapsing canyon wall. But no one had any idea just how devastating the ef-fects were, or how dire conditions were in the three riverside camp-grounds. Impounded water was overtaking them. The quake and landslides had destroyed the only road in and out of the canyon.

Meanwhile, the 200 survivors well enough to walk made their way to higher ground, eventually assembling en masse in a meadow. There, the Painter twins had their first meal in nearly twenty-four hours. Myrtle and Ray lay on a mattress inside a camper. One survivor tasked Carole with tending to them. She was still wearing just a coat. At one point, Ray called for the twins. They found their parents lying soaked in their own blood. Myrtle begged them not to come inside—she didn't want them to see her like that. Ray told the girls their mother might die. That it was up to them to get help. But how?

None of the survivors had any way of notifying the outside world about their plight. That same landslide had buried telephone lines, rendering contact with the outside world impossible. Ham radio operators rushing to the scene found that even they were useless—the steep mountain walls foiled their relay. The elevation of the canyon made flight difficult for the small helicopters normally used in the region. Desperate, the twelve-year-old twins, Anita and Anne, spelled out "SOS" with a bag of pancake flour. It was all they could think to do.

Hours later, the first helicopter arrived. Myrtle and Ray Painter were the first to be evacuated. Rescuers pinned their names, written in pencil on scraps of paper, to their shirts, and then loaded them into an army-issue helicopter. The couple's first stop was the West Yellowstone airport, where they lay on bales of straw until an old B-18 plane could deliver them to Bozeman.

There, the Painters and other survivors arrived to find a hospital ill equipped to deal with the magnitude of the tragedy: a shortage of staff; low inventories of blood, antibiotics, surgical instruments, and bandages. Stable patients were farmed out to smaller facilities. Meanwhile, overworked doctors struggled to treat conditions they'd never before seen, including gas gangrene—commonly thought to be restricted to battlefields. It's not clear whether Myrtle Painter contracted gangrene or another major infection. Whichever it was, her weakened body was no match for it. She died two days after the quake.

Back in the canyon, army reserve officers teamed up with civil defense rescue crews to begin the long search for the missing—and the dead. Mines in the region offered to send their rescue teams. Local officials didn't know whether they needed to accept: None of the campgrounds kept firm numbers of their guests, so no one knew how many individuals were actually missing. When asked, some survivors thought the slide might have buried a half dozen campers; others thought that number was probably in the hundreds. Eyewitness accounts were no less sketchy: A car carrying what those on the scene could only guess to be "four or five people" plunged into the lake and was never seen again. But witnesses couldn't say whether the occupants had managed to escape. A woman living alone was spied in her cabin as it slid into the raging river; no one knew what became of her, either. A family who had just returned from a late fishing expedition reported seeing ten pop-up trailer campers ripped into the water by the approaching tidal wave—they weren't sure if anyone was in them or not. Another man watched as a half dozen cars were buried in rock; he could only hope they were empty.

Rescuers kept looking anyway. As days went on, bulldozers and earthmovers began to dig into the slide, but they found it was just too immense. With little else to do, search and rescue teams began to walk atop its expanse, looking for clues that might reveal who lay underneath. One of them encountered Princess, the Painters' dog, wandering the wreckage and still wearing the secondhand collar. That prompted a series of frantic calls to Salt Lake City: Had the dog's human been reported missing? The owners of the deceased dog were flummoxed: Why would anyone think a long-dead pet was walking around a landslide? Or that they were underneath it?

Meanwhile, family members of actual campers chartered private planes to look for their loved ones. At the nearby airfield, the chaos created by so many aircraft attempting either to evacuate or to arrive with recovery volunteers forced Montana's state civil defense director to close the runway. Survivors began to accumulate there with no other place to go. Eventually, Montana State University offered to house them in their dormitories. Calls from across the continent began flooding into the Red Cross from worried family members and friends who still hadn't heard from their loved ones. They provided descriptions and photos, sales receipts and anything else that might prove helpful.

It took months to sort out details and identities. To this day, some questions still remain. The official death toll from the quake is twenty-eight, most of whom were victims of the landslide. It's hard to say for certain, though: None of those bodies have ever been found.

All told, the Hebgen Lake earthquake and subsequent landslide caused about $11 million in damages—over $70 million today. That price tag doesn't include residual damage, like a season's loss of vacationers. With the West Yellowstone train station condemned and its hotels deemed uninhabitable, the tourist town suffered mightily that year. Inside the park itself, staff tried their best to put on a brave face. Cafeterias began offering the "quake special"—a tossed salad mixed with hamburger—but few visitors took them up on the deal. Many others never made it at all—they canceled their reservations without asking for a refund.

Residents in six different states reported feeling tremors associated with the Hebgen quake. It left fissures in the ground four feet wide and ten feet deep. It flooded some springs and dried up others as far away as Hawaii and Puerto Rico. Rubble from the quake blocked four of Yellowstone's major thoroughfares, closed the park's west entrance, and forced the evacuation of the National Park Service office there. It damaged regional schools beyond repair and broke the seismograph at Regis University in Colorado.

Still, the Hebgen quake was by no means one of the planet's strongest. Earthquake magnitude is measured exponentially using what seismologists call the moment magnitude scale, abbreviated by the letter "M." (Another pro tip: Don't ever say "Richter scale" to a seismologist. There's never actually been such a thing, just a set of logarithms developed by a guy who shares the name and wanted to gauge waves produced by California quakes.) An M 6.0 quake is ten times stronger than an M 5.0 quake and releases thirty-one times more energy. In other words, that 5.0 quake will release the energy equivalent of 4 million pounds of dynamite. The 6.0 quake is more like 4 billion pounds—making its energy release on par with the Little Boy nuclear bomb. The quake that rocked Hebgen measured 7.3, making it one of about twenty major earthquakes (defined as M 7.0 to 7.9) experienced by our planet each year. By comparison, the infamous earthquake that rocked San Francisco in 1906 is estimated to have been a magnitude 7.7—or five times as big. The strongest earthquake on record occurred in Chile a year after Hebgen and measured 9.5—1000 times stronger than the one that killed Myrtle Painter.

It probably goes without saying that, while tragic, the Hebgen Lake quake was nowhere near the mostly deadly (that grim title goes to the 1556 quake that killed an estimated 830,000 people in China). It also wasn't even close to the most expensive: The 2011 earthquake in Tohoku, Japan, came with a price tag of over $235 billion. Nevertheless, Hebgen Lake demonstrates how seemingly powerless we were in the face of even a modest earthquake. That's probably even truer today.

COFFEE IN SALT LAKE

Today, the Painter twins are seventy. Carole is seventy-four. She's never recovered from the quake and its aftermath: The night spent tending to her parents cut too deep into her psyche. Anne doesn't like to think about the quake or what happened there. It's hard for Anita, too. She first visited the site in 1996—it took her that long to summon the courage. Being there still terrifies her. But she goes back, because she doesn't want people to forget. She's written two books about the quake and its victims. She gives talks and signings at the Hebgen Lake Visitors Center under her married name, Anita Thon. There, new visitors usually begin by watching a short film about the quake. Anita says it's not uncommon for them to come out crying or wanting to hug her. That means a lot to her.

"I'm touched that they're touched," she tells me.

Anita is pretty sure she has post-traumatic stress disorder. To this day, she can't really spend time by the side of a mountain or near a river. She still has nightmares, too. Really bad ones. A few years back, she babysat her grandkids while her son and his wife took a quick vacation to Hawaii. They live in the Wasatch Mountains. She woke up in the middle of the night to a full moon. It was too much. Flashbacks also stop her in her tracks when she least expects it—like when she smells gasoline from a neighbor's lawnmower. Mostly, though, she misses her mom.

"I loved everything about her," says Anita today. "She was so selfless. Everything she did was for her family. She would go without so that we had what we needed. I can still picture her face. She would have been such a wonderful grandmother."

Anita has eight grandchildren of her own now: four boys and four girls. They, along with the rest of her family, traveled to Yellowstone for Anita's first book signing. She took them to the site of the quake, too. Afterwards, Anita's daughter-in-law asked the kids what they liked the most about the trip. For one of Anita's young grandsons, the answer was complex: "Seeing the slide," he said, "and being glad that Grandma didn't die."

She says she feels a little guilty about that sometimes—that she cheated death. But she also derives a lot of solace from her religious faith. She knows she's been given a second chance. And she figures she owes God something at this point. The books and the talks are one way she's trying to make good on that debt.

"I try to be the voice for the people who died," she says, "so that they're not forgotten." She's researched the story of each and every one of them. Along the way, she's made friends with their families. She didn't expect that. Nor did she expect to meet the nurse who tended to her mom in the Bozeman hospital or the smoke jumpers who were first on the scene. They all remembered the Painters—especially Myrtle.

Anita and her husband reside in Utah now. She works at a medical clinic—she thinks maybe that's a way to pay back, too. Their house is also not far from the Wasatch Fault. Two hundred and forty miles long, it's one of the country's biggest. The fault itself is made up of several segments. Geologists think each segment is capable of its own magnitude 7.5 earthquake. The last rupture there was 1400 years ago.

"We're due to have a really, really big one," Anita says. "The thought of it just terrifies me—that my family members might have to go through something like I did. It happens without warning, while you're just sitting there. That's what is so scary about it."

Bob Smith agrees. He visited Hebgen Lake not long after the quake; it was that experience, more than any other, that set the course of his career.

Smith began his Yellowstone seismological work in 1973, the same year the US Geological Survey established the Yellowstone Seismic Network amid growing concerns that the caldera might be more active than previously thought. The agency tried to shut down the network in 1980 in an attempt to reallocate funds toward more populated places like San Francisco. That made Smith mad. And he said as much. The USGS told him that if he thought the network was that important, he should run it himself. He agreed, but only if they promised him better equipment.

Seismographs are delicate, finicky devices. The first one was invented by Chinese mathematician Zhang Heng in 132 CE. Zhang was a renaissance man long before the Renaissance: He wrote rhapsodies and poems on subjects ranging from the glory of a metropolis to fertile fields to the agony of being spurned by a lover. He served as chief astronomer to the Han dynasty, for which he was paid 600 bushels of grain (proving that underfunding state science is a tradition that goes way, way back). And, of course, he was keenly interested in earthquakes. China endured a lot of them, even then. And in an era when only the nobility were allowed to use horse-drawn conveyances (Zhang, incidentally, was also prefect of the majors for official carriages under the Ministry of Guards), getting aid to people in earthquake-stricken regions was a long, laborious process that could take weeks even after word of the quake eventually reached the capital.

Zhang's invention was built with this problem in mind. Shaped like a large urn and about six feet in diameter, his "earthquake weathercock" was embossed with eight dragons evenly spaced around the exterior of the vessel to represent the four cardinal directions and intermediate points between. Inside the mouth of each dragon rested a bronze ball. Any seismic activity would knock one or more of these balls out of the mouth of the dragon and into the upturned mouth of a matching bronze toad set on the floor below. The sound of the ball hitting the toad would alert a watchman, who could draw some basic conclusions about the strength and, maybe more importantly, the location of an earthquake based on which balls had moved and how loudly they had clanked into the toad's mouth.

It was an imprecise tool for sure, but it was also the best one the planet had for at least 400 years—and maybe even much longer. In the early 1700s, French physicist Jean de Hautefeuille tried studying earthquakes by filling a bowl with mercury, figuring he could track the location and strength of tremors based on how much of the highly toxic element sloshed out. His basic premise was that tiny spills would precede a much larger quake, and folks would get out of the room before a really big one sent a wave of mercury to contaminate everything and everyone around it. We've lost any evidence about how effective Hautefeuille's tool was, which may well be because its users all died of mercury poisoning. (Hautefeuille himself went on to invent the internal combustion engine, and so—understandably—his foray into seismology has been mostly overlooked by historians.)

Italian contemporaries of Hautefeuille tried their own hand at earthquake measurement. For their part, they mostly experimented with a series of pendulums that would swing in response to any tremor. In 1751, Father Andrea Bina did one better and attached a tip to the base of the pendulum, then positioned the whole device above a shallow tray of sand. Now, any tremor could be not just observed, but also recorded. And with that, the first seismogram was born.

In time, pendulum tips were replaced with needles capable of recording around the clock on rolls of paper. Today, most seismographs relay their information not as printed lines but, instead, as gigabytes of data sent through the Internet to powerful computers. Nevertheless, the basic technology hasn't changed all that much. Today, seismographs still use a weight that swings freely within a secure compartment and a base rooted on—or in—the earth. During an earthquake, seismic waves shake the base and the compartment, but the weight remains unaffected. The distance between it and the shaking compartment registers as a type of line graph known as a seismogram: the faster the shaking, the shorter the lines; the slower the shaking, the longer the lines.

Seismometers have always been expensive and in short supply. And so seismologists have long since had to rely on other methods as well. During the nineteenth century, Irish civil engineer Robert Mallet—widely considered one of the first earthquake scholars—did the lion's share of his work by studying damage to built structures like buildings and foundations. Japanese researchers at that time relied mostly upon questionnaires sent out to survivors of a quake. Crowd sourcing was used frequently throughout America and Europe, too—particularly after it was found that human observations tended to be more reliable than early seismometers. Austria had one of the most robust networks of reporters, most of whom were amateurs with little to no training. As one Austrian seismologist observed, in "probably no other field is the researcher so completely dependent on the help of the non-geologist, and nowhere is the observation of each individual of such high value as with earthquakes."

These networks presented their own problems, most notably the inconsistency not just from one reporter to another, but also between one country and another. As Deborah Coen notes in *The*

Earthquake Observers: Disaster Science from Lisbon to Richter, degrees of intensity meant very different things among European countries:

> Degree 1 required that the shock be "determined by a seismologist" in Italian, but "determined by a practiced observer" in French and German. Degree 2 specified that the shock be felt by "few people at rest" in French, but there was no comparable criterion in Italian. Degree 3 required that the shock be reported in newspapers and perceived "by people who are not concerned with seismology" in Italian but not in French. Degree 6 specified in French that "a few fearful people leave their houses"; in Italian that people "exited due to fear or prudence"; the Italian alone further specified that in describing the shock people would mention "that fortunately there were no damages."

Here in the United States, earthquake study was no less subjective—or sparse. Initially, this pursuit fell to the fledgling meteorological service (the precursor to our National Weather Service), and even funding for that was in short supply. Our nation didn't even get our first seismometers until 1885, when E. S. Holden, an astronomer who had just become president of the University of California, purchased two of them on his own dime. He installed them at the university a couple of years later. Holden insisted on the seismographs not because of any inherent interest in ground shaking but, rather, because Berkeley's frequent tremors disrupted the sensitive calibration of his telescopes. Maintaining a register of all earthquakes, he figured, would make it easier to keep his instruments on the level.

When Holden's devices provided crucial information about the devastating 1906 San Francisco earthquake, scholars became a little more interested. Leading the charge were a team of Jesuit scientists in the Midwest, who persuaded the Catholic Church to purchase sixteen seismometers and distribute them to Jesuit universities, thereby creating the nation's first seismic network. By 1930, Jesuits had a worldwide seismological network consisting of fifty-four stations that spread from Manila to Madagascar to Memphis. A weak bill came before Congress in 1911 to create a similar federal network, but it was rejected by both the House and the Senate.

Even the seismometers we had could be unreliable. The devices are sensitive to temperature and moisture, and they shouldn't be disturbed, unless you need to replace their batteries or needles or one of the other components requiring constant maintenance.

That's one of multiple reasons Bob Smith's network met with opposition. After all, there are few worse places for these kinds of considerations than the Rocky Mountains, where winter lasts seven months of the year and can bring with it ten feet of snow or more. When his network was formed, Smith insisted on concrete vaults three feet wide and several feet deep for each of his devices. They had to be waterproof. Insulated. They'd need redundant systems. To power them, he installed Yellowstone's first solar panels.

And then he encountered his biggest challenge of all: How do you get data from a snowbound seismograph in Yellowstone down to a lab in Salt Lake City? Those same mountains that prevented ham radio operators from relaying information about the Hebgen Lake survivors made similar relays all but impossible for Smith. He'd need to get up and over them somehow.

Just a few miles south of Hebgen Lake rests the Sawtelle Mountain. Atop it sits an enormous gray sphere surrounded by metal scaffolding. It's a Federal Aviation Administration radar relay device used to transmit the location of planes throughout the region.

"These systems are all linked with modern highly reliable electronics and power," says Smith. "And they have hundreds—even thousands—of channels. So I went to the FAA and asked if we could borrow a couple of them."

They said yes. Today, data from twenty-seven Yellowstone seismic stations flows from that radar relay into the air traffic control tower at the Salt Lake City International Airport. From there, it is pushed through standard telephone lines into a modem located at Smith's lab at the University of Utah.

That in itself is dicey. In 1983, the Borah Peak earthquake, centered on a fault in southern Idaho, shattered ground surfaces and created new scarps in the region. It also sent the Salt Lake City airport tower swaying.

"I joked with the tower that they had the data before anyone else," says Smith. Turns out workers there didn't think that was quite as funny as their seismologist friend did.

The Salt Lake City International Airport is built on unstable, water-saturated soil. When earthquake waves pass through this type of mixture, the force of the vibrations causes that soil to act like a liquid. What you thought was stable ground becomes a lot more like a rapidly melting milk shake. And that can be a major problem for everything—and everyone—resting atop it. The term "liquefaction" was first used in a geological sense in 1918, after the spectacular failure of the Calaveras Dam in California. Eyewitnesses later described the banks surrounding the dam as undulating like hurricane-swept waves before becoming a cataract of mud.

When the devastating M 9.2 earthquake rocked Alaska in 1964, much of the soil there also liquefied, prompting landslides and tumbling buildings. In Valdez, thirty-one people died after the ferry dock on which they were standing slid into the water. An Anchorage subdivision was decimated when the soil there liquefied, causing houses to bob like tiny boats, some moving hundreds of yards and colliding into one another. At the Anchorage International Airport, liquefaction was responsible for a massive 20,000-barrel fuel spill. It also collapsed the control tower, killing one worker inside.

More recently, liquefaction was responsible for the equivalent of millions of US dollars of damage in the 2011 earthquake centered on northeastern Japan. It's worth a quick Google search to turn up videos of the liquefaction in process: sidewalks sway and blister, cracks open and close, and a massive amount of muddy water is forced to the surface.

That kind of scenario is entirely within the realm of possibility in Salt Lake City. It—and 75 percent of Utah's population—lies on the Wasatch Fault Zone, the source of most of the seismicity in the area.

The international airport there is built on heavily silted, sandy soil, and its proximity to the Great Salt Lake means it's already saturated with water. Emergency managers there anticipate that a strong earthquake could produce liquefaction results very similar to those of the Alaska quake. Just a few years ago, the state was scoping new sites for its Draper prison, a supermax facility that has hosted such notorious killers as Ted Bundy and James Rodgers, who was executed by firing squad (and who famously made his last request a bulletproof vest, thereby proving that gallows humor isn't limited to hangings). Two of the potential sites for this new prison were just a stone's throw from the airport. *The Salt Lake Tribune* asked geophysicist Jim Pechmann to assess both sites. Pechmann, a colleague of Bob Smith's at the University of Utah, specializes in seismic hazards in the state, particularly as those hazards relate to ground motion around Salt Lake and the unique crustal formations that lie below the valley. His conclusion? The first of the two sites would probably see some flooding in a big earthquake event—most likely caused by dike failure or the flow of water under liquefying soil. The other could face a twenty-foot tsunami. According to Pechmann, the likelihood of either event is based on not just the magnitude of a future earthquake, but also the levels of the highly variable Great Salt Lake, which is currently at historic lows but regularly cycles back to higher levels.

As for the airport itself, a report completed by FEMA and regional emergency managers concluded that, while the runways themselves would probably fare pretty well, airport buildings would most like see costly damage, particularly structures like control towers, hangars, and terminal buildings. This damage, the report found, "may range from broken windows and wall cracks to partial building collapse. Fuel lines and tanks may be damaged, possibly resulting in fires." The city has launched a $2.9 billion construction process to ensure a seismically safe terminal by 2020.

Smith isn't sure what he'd do if a quake struck in the meantime.

But more importantly, he says, damage at the airport would be just the start of the area's problems. A few years ago, the Federal Emergency Management Agency commissioned the Earthquake Engineering Research Institute, an international consortium of geophysicists, engineers, architects, and public planners, to predict what an earthquake of magnitude 7.0 or higher would do to Salt Lake City. It is not an unreasonable scenario. Large earthquakes have occurred on one of the sections of the Wasatch Fault about once every 300 years. The section that runs through Salt Lake City appears to experience an M 7.0 quake every 1300 to 1500 years. The last one was 1400 years ago, making what the authors of the study called "a game of Russian Roulette" for the city.

So what would happen if and when that earthquake were to occur? The predictions are chilling: Casualties on the order of 2000 to 2500 people. Another 9000 or so badly injured. Available hospital beds reduced to a few thousand. The quake itself would create over 20 million tons of debris—820,000 truckloads' worth. It would fall in the form of bricks and wood and concrete and steel. And as it did, it would take out water, gas, and sewage lines, as well as entire electrical systems. Over 80,000 households—some 200,000 people—would be rendered homeless. Thousands of others would have to wait until building inspectors okayed their homes for rehabitation. It'd take 3400 inspectors working around the clock at least thirty days to certify those homes for occupancy. Where would the families stay in the meantime? How would the inspectors get into a city whose highways and runways had crumbled? Whose tracks had been knocked so askew they could no longer accommodate trains? How would the city get its dead and injured *out*?

Meanwhile, landslides would become avalanches at the city's big ski resorts, like Snowbird, Alta, and Solitude. With only one narrow mountain road leading in and out of that area, skiers could be trapped there for some time, without any utilities. The oil refineries surrounding the city would be at huge risk of either a catastrophic oil spill or, even worse, a massive blaze raging unchecked.

Even after rescuers began to find their way into the city after recovery began, residents there would still be without working sewer systems. They'd go weeks without utilities. Of the houses that could be saved, most would have electricity within ninety days of the quake, but on that same day, some 332,800 homes—likely containing about a million people—still wouldn't have potable water. What would these people do? How many of them would die from not having enough clean water?

These are the questions asked in the report. "Utah is NOT prepared for a major Wasatch fault earthquake," it concludes. Those capital letters are theirs, not mine.

It's a terrifying, apocalyptic scenario. But it doesn't have to be that way. The Wasatch Fault is going to go off. There's a pretty good chance it'll happen in our lifetime. We can't stop it. But we can plan for it. We can create resiliency plans and policies that end in a very different scenario: one with far fewer lives lost.

I stopped by Salt Lake City after talking with Smith, Pechmann, and Anita Thon. It is a stunningly beautiful city: a blend of modern steel and glass, along with temples and tabernacles that might have come straight out of Rome. They, along with brick and sandstone buildings, all sit tucked beside a dramatic mountain range—steep snowcapped mountains that rise out of the city on what they call the Wasatch Front. That front has been created by the eponymously named fault. So, too, has the valley in which the city sits. Over the course of millions of years, it has crept and strained, building some of our nation's most foreboding and beautiful geography.

Salt Lake City is also quite possibly America's friendliest city. As I strolled through its historic quarter, a young bride and groom were posing for photographs on the enormous campus of the Church of Jesus Christ of Latter-day Saints. At the photographer's suggestion, the bride hopped up onto a little half-wall surrounding one of the older buildings and wobbled unsteadily in her heels. "Gosh," said a security guard, "I'm really sorry, but I have to ask you to get down." "Gosh," she replied, "I'm really sorry. I should have known better."

As I watched this incredibly polite transaction take place, a woman about Anita's age stopped by. She was dressed in business attire and wore a name tag identifying her as a Mormon greeter. She asked what had brought me to the city. So I told her. A second later, she was on the phone with her husband. He's a structural engineer. Whenever there's an earthquake in the world, the church sends him to inspect their temples. When the 2010 earthquake struck Haiti, he teamed up with the Catholic Church there to create safe spaces for refugees. Here in Salt Lake City, the LDS Church spent untold amounts of money seismically retrofitting their buildings, though some of the more literal believers in the church say that was wasted money: *The Book of Mormon*, they point out, insists that an earthquake will be a sign of the Second Coming of the Lord, and casualties will be reserved for the unbelieving. Church leaders decided they needed more pragmatic assurances, including reinforcing all forty-four stone pillars in their tabernacle and securing its legendary pipe organ.

The fault zone responsible for these kinds of measures lies halfway across town, at the base of the Wasatch Mountains. When the Mormons settled Salt Lake in 1847, much of that area was craggy desert plains. It remained that way for nearly a century, until the post–World War II housing boom saw the construction of hundreds of modest ranch homes. Today, they are intermingled with strip malls and chain restaurants.

I find a clear sign of a scarp just behind the Kokopelli's Koffee there. The scarp runs alongside the back of the shop and a few small homes before disappearing into the rock. After walking it as far as I can, I go inside for a cup of coffee. There, Ani the barista is wiling away her time on a slow afternoon. She wears an oversized black T-shirt emblazoned with wolves and a large crystal pendant hung around her neck with twine. Her hair, which is dyed the color of strawberry Kool-Aid, is tucked beneath a beret she crocheted. We chat. She tells me she graduated from high school two years ago—that the job here is her stopgap until she can get up enough money for college. She thinks maybe she'd like to be a teacher or a nurse—she's not sure which. I ask her if she knows about the Wasatch Fault. She has no idea. I point out the ancient scarp on the edge of her rear parking lot. She is incredulous.

We talk about the city's plan to build a new airport terminal, which seismic design leaders there hope will keep people safe. We talk about the new prison, too. As it turns out, the state selected a third location, just northwest of the airport. Its official website crowed with the headline STATE BUYS LAND FOR NEW CORRECTIONAL FACILITY FOR MILLIONS LESS THAN ANTICIPATED. But that, says Pechmann, is an example of what happens when you don't take full-cost accounting into consideration. This new site, he explains, will require at least three to five feet of fill to raise the prison itself above known flood levels. And that fill will probably only exacerbate any liquefaction hazards at the site, too. "It might be possible to build a safe prison facility at this site," Pechmann says, "but to do so will require some excellent engineering and a lot of money."

Initially, the state budgeted $450 million for the project. A city council representative told Pechmann the total cost would probably be closer to $1 billion. They're not sure where that extra money will come from. A not insignificant amount of that overrun will be because of seismic concerns. And Pechmann says the possibility of seismically induced flooding is still real. He points to the Orleans Parish Prison, where water contaminated by sewage rose to prisoners' chests as Katrina floodwaters inundated much of southern Louisiana. Other inmates reported going four days without food and water, and female inmates as young as thirteen reportedly were forced into group cells populated by male offenders.

It's an extreme story, for sure, but it's a powerful example of the ripple effects prompted by natural disaster. And it's not a terrible analogy for what Salt Lake could look like after a major seismic event.

"How could I not know any of this?" Ani asks me at Kokopelli's Koffee.

I don't have an answer. I keep thinking about that campsite back in Hebgen Lake—about what happens when the ground drops twenty feet in a moment. Or, even worse, when it destroys a family. The chalkboard next to Ani's counter lists the shop's special drinks. I suggest maybe they need another one to acknowledge the scarp. She agrees, and we spend the next hour working out the details. In the end, we come up with a winner: two shots of espresso, mixed with brown sugar. On top, an angry fault drawn in chocolate syrup cuts across the frothed milk. We name it "The Big 7.5," and Ani promises to serve it with a shaky hand. The next time you're in town, be sure to try one: It'll rock your world.

OUR FLOATING WORLD

In December 1908, Frank Bursley Taylor boarded a train for Baltimore. With a Roman nose, well-groomed goatee, cravat, and tiepin, he looked every bit the social elite his pedigree had made him. As a young man, Taylor had matriculated at Harvard, hoping to study geology. A series of vague medical issues prompted a physician to urge that he abandon the halls of learning in favor of a more robust outdoor experience. And so, upon returning to his native Fort Wayne, Indiana, Taylor set out for repeated surveys around the Great Lakes, first in the care of a personal physician (who also acted as a chauffeur) and then in the company of his new wife (who acted as both physician and chauffeur). Most of his study focused on glaciation in the upper Midwest. But along the way, Taylor developed a secondary interest in mountain formation as well. And through this inquiry, he arrived at the most audacious of theories: Mountains like the Himalayas, he determined, could only be formed by the violent meeting of crustal plates. That meant the surface of the earth was in constant motion—forming and destroying continents, pushing them here and there.

In an era when scholars believed the planet and everything on it was fixed, Taylor's idea was no less heretical than any proposed by Copernicus. And so it must have been with some trepidation that Taylor stepped onto the sleeper car bound for Baltimore. There, he intended to propose his theory at the sixtieth meeting of the American Association for the Advancement of Science, the largest such conference of its kind, and one that annually drew thousands of scholars and professors from around the country.

Taylor was scheduled to present his paper before the Geological Society of America, a subset of the AAAS. Their meeting received next to no attention from either local media or AAAS promotional materials. It's impossible even to say where it took place: The conference program lists the site only as "Johns Hopkins Geology Lab," and not even today's official historian for the university was able to tell me where that lab was located. ("It's a dead end," he apologized after much searching. "You'll have to write around it.")

What we do know is that Taylor's paper, entitled "Bearing of the Tertiary Mountain Belt on the Origin of the Earth's Plan," was the last of the morning on Tuesday, December 29, 1908—the first full day of the conference. In his talk, Taylor spelled out his hypothesis: namely, that the continents were not fixed but, rather, showed evidence of creep over time. Alaska, he theorized, had once been part of Asia; South America and Africa may well have been connected as recently as 60 million years ago.

Taylor was not the first to suggest as much. Abraham Ortelius, a sixteenth-century Flemish cartographer, came to a similar conclusion while working on the first world atlas. Francis Bacon asserted the same after returning from oceanic voyages. Antonio Snider-Pellegrini, a nineteenth-century geologist, proposed a comparable theory in his book, *The Creation and Its Mysteries Unveiled*. It even included a map that detailed a supercontinent that would later turn out to be remarkably accurate.

By the time Taylor took the lectern in Baltimore, most of his intellectual predecessors had been long forgotten—their theories quickly discarded as unsubstantiated ramblings. And as it turned out, our Indiana don's paper had no more sway than those earlier pronouncements. Taylor would recall later that it prompted much "merriment in the meeting and apparently no serious thinking." Instead, the session dismissed for lunch and the subject was never again raised. Proceedings from the conference, published a few months later, made no mention of Taylor's theory. Undeterred, he published a longer version of his paper in the *GSA Bulletin*, one of the most widely read publications in the field. It, too, received no attention.

The same couldn't be said for his intellectual rival, Alfred Wegener. Like Taylor, Wegener was also a kind of outsider in the geological world. But whereas Taylor contented himself with tours of the Great Lakes in the careful company of a chaperone, Wegener was more of an Indiana Jones prototype: a hot air balloon record holder and Arctic explorer who occasionally appeared in his office at Germany's University of Marburg, where he had an appointment in the meteorology department but was mostly on leave.

In 1912, a colleague of Wegener's invited him to present at a regional meeting of the German Geological Society. There, Wegener introduced his idea of a single supercontinent that had eventually broken into the continents now present on the earth. The only way this could happen, he maintained, was if these enormous landmasses were capable of moving. To explain how this might be, Wegener hypothesized that continents behave like icebergs floating above a liquid sphere. He published his theory in book form in 1915 under the title *The Origin of Continents and Oceans*.

Wegener's ideas, which came to be known as mobilism, flew in the face of accepted beliefs of the time, which held that any change to the earth's topography was the result of swelling and contracting, not plate movement. And while World War I would delay dissemination in English of Wegener's theory by several years, the scientific community nevertheless had plenty of vitriol saved up for its reception: "The delirious ravings of people with bad cases of moving crust disease and wandering pole plague," wrote one scholar. "Germanic pseudo science," insisted another. "Utter damn rot," penned a third. But it was the written complaints of Rollin T. Chamberlin, president of the American Philosophical Society, that really summed up the problem: "If we are to believe Wegener's hypothesis we must forget everything which has been learned in the last 70 years and start over again."

No one in the geological community wanted to do that. As late as the 1940s, those in the field commonly referred to mobilism as "a fairy tale." It might have died out completely, had it not been for a small group of paleontologists and paleobotanists who relied upon the idea of continental drift to explain how it was that identical fossils could be found in places like the west coast of Africa and the east coast of South America or, even more vexing, in Antarctica as well as India.

Then, in the 1950s, scientists began to map the ocean floors using technology developed during World War II to locate enemy assets. Two curious facts came out of that mapping: first, that an enormous underwater mountain range known as the Mid-Ocean Ridge winds its way across the floors of both the Pacific and Atlantic Oceans and, second, that distinct magnetic strips emanate from both sides of that ridge. A survey of the ridge corridor also found intense seismic activity there, along with evidence of regular earthquakes. Scientists began to speculate about whether these ridges might actually be underwater volcanoes: cracks where magma could rise to the surface and create new crust. If that were true, the magma would have to push out the existing crust in order to make room. The continents would shift.

It would take the fifteen-year expedition of the *Glomar Challenger* to prove as much. Named after the nineteenth-century HMS *Challenger*, which distinguished itself as not only the first vessel devoted wholly to research under Her Majesty's service but also the first to map much of the ocean floor, the *Glomar Challenger* was the crown jewel of the Deep Sea Drilling Project, the first international scientific drilling program. The *Challenger* was a behemoth of a steel-hulled ship, complete with a forty-five-meter drilling platform resembling the scaffolding that stood atop early oil wells. A crew of about seventy worked around the clock to take coring samples from the seafloor and were often at sea for as long as ninety days at a stretch.

One of their primary missions was to confirm the age of the seafloor. If continental drift theories were accurate, core samples taken near the Mid-Ocean Ridge ought to be significantly younger than cores taken elsewhere.

We now believe that the earth itself is around 4.5 billion years old. The road to this hypothesis began in the 1950s, when geologists used radiometric dating to determine the age of rock around the world (another boon from the nuclear arms race). They found tremendous variation—some places like Quebec, believed to be one of the oldest land formations on the planet, contain rocks that date back over 4 billion years; parts of California, on the other hand, are believed to be between 65 and 200 million years old, and even younger in some places. There is no obvious order to this dating on many continents; maps drawn based on the age of rock are interlocking swirls of different colors designating different eons and epochs. If continental drift theories were accurate, however, the seafloor ought to be a different story: tidy lines of younger and younger rock as you approach the ridge.

To find out, the *Challenger* departed Dakar, Senegal, in early December 1968. The crew began drilling a series of boreholes off the coast of Sierra Leone and worked their way west, toward the Mid-Atlantic Ridge. Each core they brought up was younger than the one before. Late on Christmas Eve that same year, they arrived at their last site—just shy of the ridge itself. The next day, they brought up the newest core yet: It was just 7 million years old. Here, finally, was indisputable proof that the seafloor was spreading. When news reached the galley, the *Challenger*'s chef baked a special cake. The chief scientist broke out a stash of whiskey. And with that Yuletide celebration, the geological world has never been the same.

What began as Frank Bursley Taylor's theory of mountain formation is now known as plate tectonics ("tectonic" from the Greek for "carpenter" or "builder," and "plate" from the Old French for, well, "flat dish"). It's an essential part of earthquake science. It explains why Iceland, which is located pretty much smack-dab on the Mid-Atlantic Ridge, is being ripped in two and why Saudi Arabia is no longer part of Africa. It also explains why, as this process occurs and more and more oceanic plate material is produced, our planet doesn't just get bigger and bigger. The earth, of course, does not have an endless supply of magma. And if it only added new rock, we'd eventually outpace linebacker planets like Jupiter and Saturn in sheer size. That's not happening, of course. And that's because, just as places like the Mid-Atlantic Ridge are belching up new plate material, other places are consuming it. The hot spots below Yellowstone and Hawaii are two small places where this consumption occurs. But the bulk of this recycling occurs in subduction zones. As the oceanic plates are driven deeper and deeper, they melt, much like continental plates that pass over the hot spots. The process by which they are driven below is quite possibly the most violent motion on the entire planet: The plates tend to stick to each other or the oceanic plate meets resistance below the continental one. Strain builds and is eventually released in the form of a massive earthquake, like the 2011 one off the coast of Tohoku, Japan. The plates there moved 260 feet as a result of that single quake, and the seafloor lifted thirty feet. A similar subduction quake was responsible for the devastating tsunami that struck the Indian Ocean in 2004.

Of the ten largest earthquakes ever recorded, only one was not a subduction event. That quake, which occurred on the border of

India and Tibet in 1950, measured M 8.6 and was still plenty destructive (it killed at least 780 people and leveled villages). It occurred because of a rupture between two continental plates—what seismologists call a continental convergence. In that case, the edges of two continental plates also experience a head-on collision. Instead of a continental plate just running over an oceanic plate, the impact of the two continental plates is more like a failed game of chicken between two freight trains: The cars buckle up and out and, in the case of plates, form mountains like the Himalayas—just as Frank Bursley Taylor had theorized.

There's a third way plates meet, and that's what geologists call transform margins: This is where two plates are sliding past each other. The most famous example of this is, without a doubt, the San Andreas Fault Zone. About 800 miles long and bisecting most of California, San Andreas marks the meeting of the North American continental plate and the Pacific oceanic plate. There, instead of ducking below, the Pacific Plate is scooching against the North American one.

Seismologists are, by and large, pragmatists. If you ask them why San Andreas is a transform boundary and not a subduction boundary (like most meetings of continental and oceanic plates), they'll shrug. "It just is," they'll say. What they do know is that both the North American and the Pacific Plates are on the move. Pushed by their respective mid-oceanic ridges, they're both sliding to the northwest. The Pacific Plate, however, is sliding faster, and that's where the seismology gets really interesting (or problematic, depending upon how you look at it). The Pacific Plate makes gains on the North American one of about forty millimeters a year. That's about the rate that our fingernails grow. But it adds up quickly over geological time. If the process continues uninterrupted, Los Angeles will be sharing a zip code with San Francisco in about 20 million years.

Like subduction zones, transform movement rarely occurs smoothly. Seismologists like to explain it this way: Rub two smooth blocks of wood together and notice the friction. There's not much. Next, rub a sheet of coarse-grain sandpaper against one of those blocks and notice how much more energy you have to exert to get the same motion. For real fun, rub two coarse-grain pieces of sandpaper together—now you're getting close to what happens on a fault. To get these two sheets past each other, you have to push pretty hard. The Pacific and North American Plates are pushing past each other with even more force. Sometimes, they meet little resistance and creep past each other. Most of the time, nubs or even giant hunks of rock push up against each other, adding to the pressure and making it harder for the two edges to pass by each other. Nevertheless, the plates keep pushing against the strain. Eventually, something breaks.

That's what happened in the famous San Francisco quake of 1906. The built-up pressure was so intense that, when the two plates finally ruptured, they lurched about twenty feet in less than a minute's time and broke a 296-mile length of the San Andreas Fault (compare that to the 1989 Loma Prieta earthquake, which ruptured only about twenty-five miles of the fault).

In the interest of full scientific veracity, we should pause here and say that the term "San Andreas" means different things to different people. To most seismologists, it's a fault zone—a place where the convergence of the plates has created lots of faults and scarps and other forms of scarring. The USGS describes San Andreas this way: "a fragmented zone locked in a tectonic battle over which areas will give way, producing some of the steepest mountain ranges in the world."

That both this zone and its major fault occur on land makes it a real rarity, which is one reason it's so interesting to geologists. That it slices through some of the most densely populated counties in our country is another reason, which also makes the zone a nightmare for emergency managers. The fact that it has a track record of creating really terrible earthquakes that have disrupted the lives of lots of people lends to both its fame and its notoriety.

The public is even more obsessed with the fault than seismologists are. It is, as far as I can tell, the only plate boundary with its own Facebook fan page (at last count, over 1700 likes and 112 ratings, most which were five-star—not bad for a geological feature). It's also a pop culture icon: the name of the fictional setting for one of the most popular versions of Rockstar Games's *Grand Theft Auto* series, the source of narrative conflict in dozens of Hollywood films, and the denouement in at least as many novels.

Right, wrong, or indifferent, you can't talk about earthquakes in America without talking about San Andreas. It's the elephant in any seismological room, the reason many nonscientists are convinced California is going to fall into the ocean (spoiler alert: it's not). Still, there's a really good chance some portion of it is going to rupture in our lifetime.

Every few years, the USGS issues what they call a "rupture forecast" for California. They've consistently forecasted that the chance of a magnitude 6.7 or larger quake hitting the state in the next thirty years "remains near certainty," which they define as "greater than 99 percent." These are not odds most scientists ever get to experience—not even hurricane forecasters in Miami.

That's one big reason why the San Andreas Fault Zone has long enjoyed mythic status in the United States. In 1965, the US Department of the Interior designated it a "Registered Natural Landmark." Shortly thereafter, they posted a plaque to that effect at DeRose Vineyards: Built directly on the fault, the winery's tasting room is being torn in half at a rate of about an inch a year. A huge crack runs through the floor there. An outer wall has become so warped its doorway is unusable. Outside, concrete culverts bend in weird angles. In the last few decades, the vineyard has become a regular stop on seismology tours of the state. Winery employees have become skilled at figuring out who is there to taste wine and who is there to check out the fault.

"Khaki pants and the funny hat, and you know it's a geologist," co-owner Alphonse DeRose told *The Wall Street Journal* in 2009.

Perhaps the fault zone's biggest fan of all is astrophysicist David Lynch, who spends his downtime tracing the fault—*all of it*. In 2004, he decided on a whim to drive the entire length of the fault for what he says was his "own amusement." He quickly realized doing so was a nearly impossible task without a vanload of various types of maps and geological textbooks. Lynch is an altruistic guy, and it occurred to him that he could save the rest of us the same trouble if he just plotted the whole fault zone for us. It took him nearly two years, and the end result was *Field Guide to the San Andreas Fault*. It is a fabulous road trip book, complete with useful tools like inch-by-inch GPS coordinates of the fault, along with recommendations for pit stops and places to, in Lynch's word, "lollygag." I used the guide to follow most of the fault zone, or at least the zone from north of San Francisco to the base near the Salton Sea. At times, I even cajoled very smart (and even more accommodating) sciency friends to join me. But it didn't take more than the first couple of hours of driving down dirt roads and parking under overpasses for us to learn a very important fact: Faults are hard to see. Sure, it's pretty easy to see a scarp like the one at the Cabin Creek Campground, or to realize that one side of a canyon is a completely different type of rock than the other side (that happens a lot in California, as the zippy Pacific Plate continues to outpace the North American—and takes all of its geological features with it in the process). Seeing the actual fault, on the other hand, well, that takes a lot more than a car and a clever field guide.

Most of the faults composing the San Andreas Fault Zone are deep belowground—in some places as far down as ten miles below the surface. And so, as it turns out, if you want to actually see what the fault looks like down there, you don't go to California. Instead, you drive about 1500 miles east, to College Station, Texas.

COLD STORAGE

After the *Glomar Challenger* returned from its landmark mission, the scientists involved recognized that they needed to store their core samples somewhere safe—somewhere they could be preserved for posterity. Within a few years, these cores began to add up and were soon joined by others from around the world, each a chapter in the narrative that is our earth's history. Out of this work came the formation of a new multinational organization: the International Ocean Discovery Program. Recognizing the value of these core samples, IODP heads established three state-of-the-art repositories located in Kochi, Japan; Bremen, Germany; and College Station. That last one is housed on the campus of Texas A&M in an unassuming concrete building not far from the George Bush Presidential Library and Museum (a much nicer concrete building).

Curating the entire IODP collection is John Firth. I went to visit him because he agreed to show me cores taken from the San Andreas Fault Observatory at Depth (SAFOD) drilling project—the first of its kind in the world. Prior to the core's existence, geologists were largely limited to conducting research on faults at the surface of the earth. That's really useful for explaining past quakes (a field of study known as paleoseismology), but it doesn't tell you much of anything about what's going on in real time. The core, on the other hand, tells you quite a lot.

Every scientist doing work on subjects ranging from how transform faults work to what triggers an earthquake wants a piece of this core. It's Firth's job to sort through the hundreds of requests that come in every year. Understandably, he's a busy guy and hard to pin down.

We found a brief window in May, during a lull between when competing researchers register to request sections of the core and when they have submitted their formal proposals. Three days before our scheduled meeting, Firth emailed a warning that he might have to cancel: A migraine had laid him out flat. But he rallied the morning of, and I found him looking bright-eyed and jovial in his standard-issue academic office, perfectly cluttered with bookshelves of yellowing journals, stacks of papers, and just a few personal touches (in this case, brightly colored drawings declaring him the "best dad ever," ground coffee in a giant Mason jar, and what appeared to be the entire collected works of C. S. Lewis).

Despite the fact that it was already 90°F outside with a humidity level to match, Firth seemed utterly comfortable in a pair of black slacks and a yellow shirt in a heavy cotton weave. We shook hands and, apologizing for the accommodations, he offered me one of two broken chairs lined up next to his desk. We chatted about the Texas heat (because you can't not do that in Texas) and the oil industry (which you also can't not do when you're in Texas).

It's probably rude to ask someone who spent the weekend in bed with a splitting headache to spend his morning touring you through a 150,000-square-foot core repository. I did it anyway. And Firth obliged, which meant I also soon understood his unseasonable wardrobe. IODP keeps their holdings at a frigid 40°F (which happens to also be the same temperature as the bottom of the ocean), sealed with the kind of door you'd find on an industrial walk-in re-frigerator.

Beyond that door is nothing short of an enormous warehouse in which tens of thousands of identical cores, each about two to three inches in diameter, are sheathed in identical white tubes and stacked in shelves over ten feet high. Together, they compose an encyclope-dia of the earth's crust. And Firth is an encyclopedia of that ency-clopedia. Imagine a very large, very cold library. Now imagine the librarian who can pull any book out of those stacks by memory, no matter if you are asking for Dr. Seuss, *Doctor Zhivago*, or Dr. Living-stone. Now you have a pretty good approximation of Firth. He has committed to memory the archival proof of how our planet came to be and can summon a single core with more ease than most of us can find a matching pair of socks.

While Firth and I talked about the repository, Daniel, a lanky re-search assistant with hipster jeans and a clipped Spanish accent, checked in with us. He'd been told I was visiting and why. Firth rat-tled off from memory codes for each of the three San Andreas cores he wanted to show me, and Daniel excused himself to locate them.

In total, Texas A&M has about 200 feet of San Andreas core, each cut into cylinders about 4 inches wide and about a yard long. They're kept in their own section of the repository and stored in long pine boxes that look like old-timey coffins. Standing in the refrigerated warehouse, Firth invited me to step up on a ladder to take a look inside one or two as he pulled them out to the end of the shelf. By this point, I was also shivering, even in the jacket I'd thought to bring, and so was pretty much scared to death I'd knock over the whole lot. Firth must be accustomed to this response, be-cause he quickly packed up the samples and led me back to the lab, where we joined Daniel, who had laid out several core cross sec-tions for us to look at in detail. Each was a mess of chipped rock and oddly shaped Styrofoam placeholders, all wrapped in plastic film that looks a whole lot like kitchen cling wrap.

It's Daniel's job to mind the core. He knows every millimeter of it—what samples have been taken, where it's most fragile, how to tuck it into its plastic to make sure it remains pliant and intact. Watching him, I was reminded of my favorite baker. She named her successful sourdough starter "Lulu," and she takes it with her everywhere—on ski trips, to her summer cabin on the ocean, always lovingly enshrined in a plastic Marshmallow Fluff container and a thick knit cap to keep it safe and warm. Daniel's relationship with the core is a lot like that, which is understandable, given what's riding on it.

Overseeing administration of the core is the National Science Foundation. Judi Chester, a geology professor at Texas A&M, serves as their liaison along with Firth. The NSF has mandated that every single micron of the core be accounted for at all times. Any piece removed for research is replaced with one of those Styrofoam placeholders, each bearing the name of the successful requester who received it. It's kind of like the Hollywood Walk of Fame for fault zone studies. I knew I'd successfully geeked out on this subject when I got giddy at seeing the names of some of my favorite researchers hastily written with a Sharpie on tufts of white foam.

The cores themselves vary considerably based on where they were taken in relation to the fault. Some are slick and dark like coal; others are sandstone as bright as any desert sunset. Their texture is equally varied. Some samples appear perfectly, monochromatically uniform. Others contain textured variation created by pebbles and rock fragments. This variability increases significantly in the cores taken closest to the fault itself. Look closely, and in those samples you'll find more and more fractures—from tiny cracks to immense shatters that look like they must have been caused by a grenade or something just as powerful. In the heart of the fault, the cores are composed almost entirely of a substance geologists call fault gouge: tiny fragments of rock not unlike a fine gravel created by the friction and slow creeping of the plates as they rub against each other.

How this fragile slurry came to rest in a Texas lab is a story for the ages. Prior to the start of the SAFOD project, the team met with drilling experts in the oil industry to find out how to go about a project like this one. The oil experts had no idea—it has always been their policy to avoid drilling anywhere near a fault zone. The SAFOD team decided to improvise. Their initial plan was to drill down about six miles, but that soon proved too ambitious—and expensive. Eventually, the SAFOD team settled on a three-pronged approach: They'd begin with a straight borehole down just shy of a mile, then angle their way farther down and across the fault zone. That would give them samples from both the Pacific and Continent plates, as well as the fault itself. Along the way, they'd establish a string of instruments capable of monitoring the fault zone long after the core was removed. They consulted with NASA on how to prepare these instruments for the rigors of a life at nearly 200°F in an atmosphere filled with corrosive gases. Even with all this preparation, the instruments soon failed under the conditions. For a while, it looked like the coring project might as well: In their first attempt, the drill became stuck at depth. During a second attempt, it skittered off granite and sheared the communications lines for some of the instruments. On the third attempt, everything went as planned.

Judi Chester was there when samples actually began to appear aboveground. She says everyone was holding their breath. Chester and the other geologists on staff conducted a cursory examination of each core as it came to the surface, then labeled each side so they'd always know its spatial orientation and from where it came. They placed them all on padded pallets and shipped them to College Station in refrigerated tractor-trailers. Next, they sent them over to Texas A&M, which has a special high-resolution CT scan. And then they pretty much sat and looked at the samples for a couple of years, trying to figure out what to do next—and how not to wreck these incredibly valuable samples.

"This wasn't pliant ocean mud," says Firth today. "We had no idea how to cut it without letting it crumble to pieces." They tried precision lasers and high-pressure water jets: Neither worked. Then someone had the idea to go old-school. They wrapped the core in the same kind of aluminum sheeting used to make soda and beer cans, which served as a kind of jacket for the brittle core. Blammo: They could cut the samples with no problem.

Bigger issues arose when they tried to figure out how to parcel out samples of the core. Firth received requests for nearly 800 different samples the first year alone. Most everyone wanted the gouge. No one got it.

"It was just too precious," he says. "We needed to know that studies were repeatable, that the science was sound." And so, initially, requesters got chips and bits—tiny castoffs from gouge samples that wouldn't affect the cores themselves. In time, they began to give away larger and larger sections. But in the interest of historic preservation, the SAFOD committee has decided that at least 25 percent of any given section of the core must be preserved in perpetuity. They created a 3-D map of the core so requesters could see if what they wanted was even available.

They also created a life-size model of the section of core taken directly from the fault itself. Even this replica makes Firth's graduate assistant Daniel nervous. As John pointed out the different sections and their significance to me, he spun the model from end to end, pointing out the different striations and their significance. Each time Firth touched the model, Daniel flinched and took a quick deep breath.

In total, hundreds of studies on the core have been conducted by scientists around the world. They've examined the texture of the core rocks. Their chemical makeup. How waves travel through them. Their electrical conductivity and porousness. What happens when you compress them. Crush them. Heat them. Shock them. Gas them. Taunt them (I may have made up that last one). They've simulated quakes in labs and examined the microscale physics of fault slivers, all in the hopes of figuring out how earthquakes work and when they happen.

As part of that mandate to account for every micron of the core, NSF insists that every sample be returned to the IODP depository once these studies have been completed. Researchers comply, but the results are kind of comical. Inside the SAFOD "headquarters," which is actually just a low-rent office with a tattered desk and a run-of-the-mill PC, a set of metal bookshelves holds the remnants of studied core. It arrives powdered and granulated, in test tubes and tiny snack-size Ziploc baggies. It is returned on microscope slides and encased in plastic molding like the kind that might hold an ancient ant or a really rare coin. Each sample, no matter how tiny, decimated, or altered, also returns with a full account of everything that's been done to it. Some of these accounts are about as long as this book.

This is how earthquake research often happens, says Judi Chester: sometimes in fits and starts, most of the time in laborious experiments that may or may not yield any results. Still, she says, it's more than worth it. Researchers studying the SAFOD core have already learned some important truths. They've discovered that different types of rocks don't always rupture the way you'd expect. Prior to this project, researchers always assumed that a massive fault zone like San Andreas would have an equally massive rupture zone. As it turns out, San Andreas's is quite narrow—just a couple of feet wide in some places. Knowing what it looks like has profound implications in places you might not expect, like the siting of water and gas lines.

SAFOD has also revolutionized how faults are studied. When the project began, scholars speculated that it might be a one-off—almost a gimmick, really. But that's not the case. Scientists from around the world now rely upon similar projects to do much of their work. One of the largest fault drill sites currently operational is at the Nankai Trough in Japan. They're hoping to get down over three miles, into the heart of the fault itself. A similar project is about to begin in a large subduction zone off the coast of New Zealand. Best guesses are that it'll cost about $20 million and take three years to complete.

That's a bargain, says Chester. USGS estimates put the price of SAFOD around $30 million. But everyone associated with the project agrees its value far exceeds that. Talk to researchers who work with it, and you'll hear words like "precious" and "invaluable."

How valuable? I ask Chester as we walk back out into the interminable Texas heat.

"As valuable as moon rocks were in achieving our understanding of the moon," she says. "Maybe even more."

Heather Savage is one of the recipients of the SAFOD core. She's a geophysicist at Columbia's Lamont-Doherty Earth Observatory. You can get most of the way there by train, but then you have to take a special university bus from Harlem an hour or so into the pastoral loveliness of New York's Palisades region. There, some of the smartest people in the world try to figure out some of the universe's biggest mysteries on a campus that still looks very much like a weekend country estate for the elite.

Savage meets me at the observatory bus stop wearing a backpacker's rain jacket, along with jeans and running shoes. She says a lot of people still mistake her for a graduate student. I can see why.

Our first stop is the Rock and Ice Mechanics Lab. Savage spends most of her time there. More specifically, she spends most of her time in what used to be a closet in that lab. It now contains her pride and joy: a superpowerful machine capable of replicating a very small earthquake.

Savage stands about five foot six, maybe a little taller. Her machine is about as tall as she is. An enormous metal base includes a hollow column made of steel. A large tank of argon sits nearby. When she pumps it in under intense pressure, she can simulate the conditions three miles below the surface of the earth, where most of our earthquakes occur. That takes incredible force.

"Basically," she says, "this machine is a bomb."

No one is allowed inside the room when the machine is operating. The tiny space has been retrofitted with metal riveted shades that researchers must close over the windows before powering it up. The ceiling is similarly reinforced, though someone does have the office directly above.

"I don't love that," Savage admits.

Once her machine has achieved something akin to the condition under the surface of the earth, Savage can employ a hydraulic piston to force a rock sample to break or crumble. If she's started a little tear in it, she can simulate what happens on a fault, too. It's a decent approximation, she says, and pretty much the best tool we have for understanding what happens during a quake. But, Savage adds, it's a far cry from the real thing.

"These lab events are just so, so, so tiny," she cautions. "That makes them really different than what happens in the world."

That's one reason, she says, why our ability to understand earthquakes has been so slow to evolve.

Since time immemorial, we've struggled to make sense of the phenomenon. Ancient Greeks believed earthquakes were caused by subterranean winds. The Aztecs believed they were the devil himself trying to escape from deep inside the earth. The Norse believed they were caused by an enormous wolf howling in fury. Medieval philosophers alternately thought they were earth's flatulence, let loose when that air was warmed by the sun, or God's vengeance. Leonardo da Vinci capitalized on the air theory by proposing that earthquakes occurred when collapsing mountains sighed. The renaissance astronomer Johannes Kepler argued that they were provoked by "animated celestial bodies," particularly angry comets. Gottfried Wilhelm Leibniz blamed volcanoes (not a bad call). Immanuel Kant contended they were actually complex meteorological reactions involving "2 parts vitriol oil, 8 parts water, 2 parts iron." It wasn't until 1760, when British engineer John Michell speculated that earthquakes were waves of energy caused by "shifting masses of rock miles below the surface," that we began to understand the phenomenon.

Meanwhile, other seismological advances came only by way of industry and necessity: Faced with the onset of mechanized combat, British military leaders developed ways to record acoustic waves as a means of determining the location of German tanks and cannons on the battlefields of World War I. Beginning in the 1920s, the petroleum industry used similar methods to determine the location of pockets of oil. But it would take the messy end of the Second World War before US seismic studies would really get a boost.

More than anything, our nation's commitment to nuclear arms has fueled its current approach to seismological studies. In the years after V-J Day, terrifying images from Hiroshima and Nagasaki left people around the world calling for an end to nuclear proliferation. Those images had the opposite effect on the Soviet government, which, fearing a weapons gap, began intense nuclear experimentation. Murmurs that Britain and France were to follow left US officials in a panic. They responded by creating the Atomic Energy Detection System, a highly classified project to track nuclear activity around the world. Members of the AEDS team used a combination of different techniques, including radiological testing of rainwater, to locate any potential detonation. Of all these techniques, seismic monitoring quickly proved one of the most effective: Even when detonated well underground, a nuclear bomb shakes the earth enough to register on seismograms around the world.

This fact proved particularly valuable in the early months of

1959, when delegates from the most powerful nations finally met to broker a moratorium on nuclear testing. Still beleaguered by World War II and its ugly conclusion, international leaders arrived with little trust and even less confidence in the binding nature of a treaty. For a moratorium to work, they agreed, the world would have to invest in a nuclear monitoring system capable of detecting any infraction or treaty breach. Soviet representatives contended that existing seismic stations were more than sufficient for the task, but the British and US delegates disagreed.

What the US delegates failed to mention was why they were so confident in the shortcomings of existing networks. For several years, the Department of Defense had been conducting underground nuclear tests, part of two projects known as Plumbbob and Hardtack. Both projects had taught them that little of the energy from an underground nuclear explosion is actually released, and when it is, it looks pretty different from an earthquake. That made it easy to pinpoint which seismic activity was probably the result of nuclear detonation, but also meant they'd need exponentially more seismic stations if they were to detect these contained explosions. With that in mind, the US delegates demanded a global seismic network of at least 650 stations. Perhaps not surprisingly, the Soviets balked.

The two nations were still very much at loggerheads when the Hebgen Lake quake ruptured. At the time, the closest seismographs to Hebgen Lake were in Bozeman and Salt Lake City. Beyond that, you'd have to travel to Reno or California. This lack of instrumentation made it hard to say anything much more specific than that an earthquake had occurred and that it was a strong one.

It wasn't until 1961 that international leaders agreed upon the creation of the Worldwide Network of Seismograph Stations (WNSS), a linked system of standardized recording devices. In their proposal, US leaders suggested that the initial forty-seven stations be established at various points throughout the Northern Hemisphere.

Given the widely known fact that we did most of our nuclear testing in the South Pacific, this plan did little to engender friendly feelings in Russia. And so the stalemate continued. Meanwhile, the United States blithely continued its nuclear testing. On January 19, 1968, we began a new project at the Central Nevada Supplemental Test Area with the quippy codename Faultless (other than geologists, no one has a sense of humor quite like a nuclear scientist—especially one working for the government). The resulting nuclear explosion set off a 400-foot rupture at the site and may have prompted seismic activity as far away as twenty miles.

Around this time, the United States also launched a new project called Project Plowshare, an unfortunate allusion to the Old Testament book of Isaiah, which prophesized a moment when, following divine judgment, the people of Israel would "beat their swords into plowshares and their spears into pruning hooks."

Our iteration looked a little different. Project Plowshare focused on three initial undertakings each using nuclear weapons to irreparably change the landscape: The first project consisted of detonating a channel through the reef surrounding the Marshall Islands. The second involved blowing up a bomb off the coast of Alaska in order to create a new harbor, and the third detonation was supposed to create a new system of canals. A fourth, Project Gnome, involved blowing up a salt dome to "study energy production." So began a robust era of nuclear explosions in the country, each requiring its own seismic account of rumblings in the earth. The Plowshare program culminated in 1973 with Project Rio Blanco, in which three 33-kiloton bombs (each twice the strength of Little Boy, the bomb that devastated Hiroshima) were exploded near Rifle, Colorado, with the hopes that it might stimulate natural gas production. It didn't. But it did prove that these kinds of activities have a measurable seismic effect.

Nuclear testing continued at the Nevada Test Site until 1992 and did much to sophisticate how seismic networks operate in the United

States. As the Cold War reached its apex, the United States saw these networks as their best defense against a nuclear attack from Russia. A 1988 report published by the US Congress Office of Technology Assessment warned:

> For the United States, the main national security benefits derived from test limitation treaties are a result of the Soviet Union being similarly restricted. In considering agreements that bear on such vital matters as nuclear weapons development, each country usually assumes as a cautious working hypothesis that the other parties would cheat if they were sufficiently confident that they would not be caught. Verification—the process of confirming compliance and detecting violations if they occur—is therefore central to the value of any such treaty.

Seismic monitoring, argued the report, continued to be the most effective way to verify that treaty.

Knowing full well that a nuclear explosion manifests on a seismogram differently than an earthquake, the US government began to consider ways in which the Soviet Union might attempt to mask any such detonation. They decided the most likely ways were to bury it deep underground, to make it look like "a large legitimate industrial explosion," or to time the test such that it occurred with a natural earthquake. We also considered the prospect that the Soviet Union might try to test its nuclear weapons "behind the sun," but eventually concluded there might be a few logistical problems there. Regardless, we were sure they were up to something: "A country considering cheating," read the report, "would have to evaluate the risks and costs of being caught against the benefits of succeeding. A country concerned about preventing cheating has to guess the other country's values for making this decision."

In the end, we decided the Soviet Union was pretty likely to cheat. In 1987, the United States established a large seismic network in Norway to complement others already in North America. They didn't do much to catch the Soviet Union red-handed, but they did teach us a lot about seismicity.

We now know that most of the planet's earthquakes occur at tectonic boundaries like the San Andreas Fault. Most, but definitely not all. The Hebgen Lake quake was what is known as an intraplate or non-plate boundary earthquake. Any seismic activity on the Wasatch Fault would be as well. And these quakes are also capable of enormous seismic activity.

In 1886, a big one tousled Charleston, South Carolina. With an estimated magnitude of 7.5, the quake wasn't just the strongest recorded east of the Mississippi; it was also one of the most powerful in our nation's history. And for an already beleaguered city, the effects were devastating. Just fifteen years earlier, much of Charleston had been wiped out by fire. Town officials responded by mandating that new construction be limited to brick. Hardly any of those buildings survived the quake. Homes and buildings as far away as Ohio and Alabama were also damaged during the shaking, and nearly fifty miles of railroad track were rendered unusable. Residents of Boston, Chicago, and Havana all reported the shaking.

It's easy to assume events like that one are freakish anomalies. But history tells a different story.

A couple of hundred thousand years ago, an M 7.2 earthquake shook what is now New Hampshire. Just a few thousand years ago, an M 7.5 quake ruptured just off the coast of Massachusetts. And then there's New York.

Since the first western settlers arrived there, the state has witnessed 200 quakes of magnitude 2.0 or greater, making it the third most seismically active state east of the Mississippi (Tennessee and South Carolina are ranked numbers one and two, respectively). About once a century, New York has also experienced an M 5.0 quake capable of doing real damage.

The most recent one near New York City occurred in August of 1884. Centered off Long Island's Rockaway Beach, it was felt over 70,000 square miles. It also opened enormous crevices near the Brooklyn reservoir and knocked down chimneys and cracked walls in Pennsylvania and Connecticut. Police on the Brooklyn Bridge said it swayed "as if struck by a hurricane" and worried the bridge's towers would collapse. Meanwhile, residents throughout New York and New Jersey reported sounds that varied from explosions to loud rumblings, sometimes to comic effect. At the funeral of Lewis Ingler, a small group of mourners were watching as the priest began to pray. The quake cracked an enormous mirror behind the casket and knocked off a display of flowers that had been resting on top of it. When it began to shake the casket's silver handles, the mourners decided the unholy return of Lewis Ingler was more than they could take and began flinging themselves out windows and doors.

Not all stories were so light. Two people died during the quake, both allegedly of fright. Out at sea, the captain of the brig *Alice* felt a heavy lurch that threw him and his crew, followed by a shaking that lasted nearly a minute. He was certain he had hit a wreck and was taking on water.

A day after the quake, the editors of *The New York Times* sought to allay readers' fear. The quake, they said, was an unexpected fluke never to be repeated and not worth anyone's attention: "History and the researches of scientific men indicate that great seismic disturbances occur only within geographical limits that are now well defined," they wrote in an editorial. "The northeastern portion of the United States . . . is not within those limits." The editors then went on to scoff at the histrionics displayed by New York residents when confronted by the quake: "They do not stop to reason or to recall the fact that earthquakes here are harmless phenomena. They only know that the solid earth, to whose immovability they have always turned with confidence when everything else seemed transitory, uncertain, and deceptive, is trembling and in motion, and the tremor ceases long before their disturbed minds become tranquil."

That's the kind of thing that drives Columbia's Heather Savage nuts.

New York, she says, is positively vivisected by faults. Most of them fall into two groups—those running northeast and those running northwest. Combined they create a brittle grid underlying much of Manhattan.

Across town, Charles Merguerian has been studying these faults the old-fashioned way: by getting down and dirty underground. He's spent the past forty years sloshing through some of the city's muckiest places: basements and foundations, sewers and tunnels, sometimes as deep as 750 feet belowground. His tools down there consist primarily of a pair of muck boots, a bright blue hard hat, and a pickax. In public presentations, he claims he is also ably abetted by an assistant hamster named Hammie, who maintains his own website, which includes, among other things, photos of the rodent taking down Godzilla.

That's just one example why, if you were going to cast a sitcom starring two geophysicists, you'd want Savage and Merguerian to play the leading roles. Merguerian is as eccentric and flamboyant as Savage is earnest and understated. In his press materials, the former promises to arrive at lectures "fully clothed." Photos of his "lab" depict a dingy porta-john in an abandoned subway tunnel. He actively maintains an archive of vintage Chinese fireworks labels at least as extensive as his list of publications, and his professional website includes a discography of blues tunes particularly suitable for earthquakes. He calls female science writers "sweetheart" and somehow manages to do so in a way that kind of makes them like it (although they remain nevertheless somewhat embarrassed to admit it).

It's Merguerian's boots-on-the-ground approach that has provided much of the information we need to understand just what's going on underneath Gotham. By his count, Merguerian has walked the entire island of Manhattan: every street, every alley. He's been in most of the tunnels there, too. His favorite one by far is the newest water tunnel in western Queens. Over the course of 150 days, Merguerian mapped all five miles of it. And that mapping has done much to inform what we know about seismicity in New York.

Most importantly, he says, it provided the first definitive proof of just how many faults really lie below the surface there. And as the city continues to excavate its subterranean limits, Merguerian is committed to following closely behind. It's a messy business.

Down below the city, Merguerian encounters muck of every flavor and variety. He power-washes what he can and relies upon a diver's halogen flashlight and a digital camera with a very, very good flash to make up the difference. And through this process, Merguerian has found thousands of faults, some of which were big enough to alter the course of the Bronx River after the last ice age.

His is a tricky kind of detective work. Think back to that gouge at Texas A&M: The center of a fault is primarily pulverized rock. For these New York faults, that gouge was the very first thing to be swept away by passing glaciers. To do his work, then, he's primarily looking for what geologists call "offsets"—places where the types of rock don't line up with one another. That kind of irregularity shows signs of movement over time—clear evidence of a fault.

Merguerian has found a lot of them underneath New York City.

These faults, he says, do a lot to explain the geological history of Manhattan and the surrounding area. They were created millions of years ago, when what is now the East Coast was the site of a violent subduction zone not unlike those present now in the Pacific's Ring of Fire.

Each time that occurred, the land currently known as the Mid-Atlantic underwent an accordion effect as it was violently folded into itself again and again. The process created immense mountains that have eroded over time and been further scoured by glaciers. What remains is a hodgepodge of geological conditions ranging from solid bedrock to glacial till to brittle rock still bearing the cracks of the collision. And, says Merguerian, any one of them could cause an earthquake.

You don't have to follow him belowground to find these fractures. Even with all the development in our most built-up metropolis, evidence of these faults can be found everywhere—from 42nd Street to the Village. But if you want the starkest example of all, hop the 1 train at Times Square and head uptown to Harlem. Not far from where the Columbia University bus collects people for the trip to the Lamont-Doherty Earth Observatory, the subway tracks seem to pop out of the ground onto a trestle bridge before dropping back down to earth. That, however, is just an illusion. What actually happens there is that the ground drops out below the train at the site of one of New York's largest faults. It's known by geologists in the region as the Manhattanville or 125th Street Fault, and it runs all the way across the top of Central Park and, eventually, underneath Long Island City. Geologists have known about the fault since 1939, when the city undertook a massive subway mapping project, but it wasn't until recently that they confirmed its potential for a significant quake.

In our lifetimes, a series of small earthquakes have been recorded on the Manhattanville Fault including, most recently, one on October 27, 2001. Its epicenter was located around 55th and 8th—directly beneath the *original* Original Soupman restaurant, owned by restaurateur Ali Yeganeh, the inspiration for *Seinfeld*'s Soup Nazi. That fact delighted sitcom fans across the country, though few Manhattanites were in any mood to appreciate it.

The October 2001 quake itself was small—about M 2.6—but the effect on residents there was significant. Just six weeks prior, the city had been rocked by the 9/11 terrorist attacks that brought down the World Trade Center towers. The team at Lamont-Doherty has maintained a seismic network in the region since the 1970s. They registered the collapse of the first tower at M 2.1. Half an hour later, the second tower crumbled with even more force and registered M 2.3. In a city still shocked by that catastrophe, the early-morning October quake—several times greater than the collapse of either tower—jolted millions of residents awake with both reminders of the tragedy and fear of yet another attack. 9-1-1 calls overwhelmed dispatchers and first responders with reports of shaking buildings and questions about safety in the city. For seismologists, though, that little quake was less about foreign threats to our soil and more about the possibility of larger tremors to come.

Remember: The Big Apple has experienced an M 5.0 quake about every hundred years. The last one was that 1884 event. And that, says Merguerian, means the city is overdue. Just how overdue?

"Gee whiz!" He laughs when I pose this question. "That's the holy grail of seismicity, isn't it?"

He says all we can do to answer that question is "take the pulse of what's gone on in recorded history." To really have an answer, we'd need to have about ten times as much data as we do today. But from what he's seen, the faults below New York are very much alive.

"These guys are loaded," he tells me.

He says he is also concerned about new studies of a previously unknown fault zone known as the Ramapo that runs not far from the city. Savage shares his concerns. They both think it's capable of an M 6.0 quake or even higher—maybe even a 7.0. If and when, though, is really anybody's guess.

"We literally have no idea what's happening in our backyard," says Savage.

What we do know is that these quakes have the potential to do more damage than similar ones out West, mostly because they are occurring on far harder rock capable of propagating waves much farther. And because these quakes occur in places with higher population densities, these eastern events can impact a lot more people. Take the 2011 Virginia quake: Although it was only a moderate one, more Americans felt it than any other one in our nation's history.

That's the thing about the East Coast: Its earthquake hazard may be lower than that of the West Coast, but the total effect of any given quake is much higher. Disaster specialists talk about this in terms of risk, and they make sense of it with an equation that multiplies the potential hazard of an event by the cost of damage and the number of people harmed. When you take all of those factors into account, the earthquake risk in New York is much greater than, say, that in Alaska or Hawaii or even a lot of the area around the San Andreas Fault.

Merguerian has been sounding the alarm about earthquake risk in the city since the 1990s. He admits he hasn't gotten much of a response. He says that, when he first proposed the idea of seismic risk in New York City, his fellow scientists "booed and threw vegetables" at him. He volunteered his services to the city's Office of Emergency Management but says his original offer also fell on deaf ears.

"So I backed away gently and went back to academia."

Today, he says, the city isn't much more responsive, but he's getting a much better response from his peers.

He's glad for that, he says, but it's not enough. If anything, the events of 9/11, along with the devastation caused in 2012 by Superstorm Sandy, should tell us just how bad it could be there.

He and Savage agree that what makes the risk most troubling is just how little we know about it. When it comes right down to it, intraplate faults are the least understood. Some scientists think they might be caused by mantle flow deep below the earth's crust. Others think they might be related to gravitational energy. Still others think quakes occurring there might be caused by the force of the Atlantic ridge as it pushes outward. Then again, it could be because the land is springing back after being compressed thousands of years ago by glaciers (a phenomenon geologists refer to as seismic rebound).

"We just have no consciousness towards earthquakes in the eastern United States," says Merguerian. "And that's a big mistake."

We don't know a great deal more about what triggers earthquakes on plate boundaries, either. That includes *when* a fault might rupture and *how*. That "how" part is called nucleation.

Savage is currently using her sample of the San Andreas core to work on the problem. Still, she says, a lab is not at all the same thing as an actual fault. "The problem comes in when you start to think of how complex fault zones really are," she says.

There's more variation in fault zone rock then she can simulate in her sample, and she can't really re-create the all-important gouge, either.

For nearly a decade, some of her peers have been hoping to glean some answers 9400 miles away on the Whillans Ice Stream. Located in western Antarctica and roughly the area of Maryland, Whillans appears utterly unremarkable to the naked eye—a windswept expanse of snow with few distinguishing features other than the occasional research tent or Ski-Doo. Stand there, and you will just as easily believe that you've landed in a Kansas cornfield in January as you will anything else. But underneath that field of snow is actually one of six major ice streams that convey the West Antarctic Ice Sheet into the Ross Ice Shelf, where it will eventually calve off into the Ross Sea. For the most part, the Whillans Ice Stream flows so slowly—about six centimeters an hour—that that movement is imperceptible to anything other than a sophisticated GPS system. However, every day, a large section of the ice stream lurches forward about half a meter in what seismologists call a slip-stick event. This activity is similar to what happens on many of the world's faults. Even more significant is the fact that scientists know this slip-stick event will happen almost every day around high tide. There's a really good chance it'll happen every day just before low tide, too (although the ice sheet sometimes skips that event). This kind of regularity isn't seen anywhere else in the seismic world. And that makes Whillans a real mecca for earthquake research—particularly the kind of research that investigates precursory activity and what it might mean for earthquake forecasting. A team of geoscientists led by J. Paul Winberry studied this activity in December 2010 and then again the following year. They published their findings in a 2013 issue of *Geophysical Research Letters*. There, the authors argued that they had gleaned evidence of "a unique pattern of precursory slip" prior to each seismic event on Whillans, including harmonic tremors similar to the ones often seen in volcanic activity.

On a seismogram, these precursory slips can look like the EKG of a marathon runner. That kind of regularity is significant, especially if you're looking for geological patterns. But it's far from sufficient when it comes to proving that earthquakes are predictable. Scientific certainty would require researchers to find proof of these patterns over time, too. And they're not sure to what degree an ice sheet really models what would occur in a real earthquake.

That's why Heather Savage would really like to trigger one of her own. She's not unreasonable: She says she could do it in a remote place, maybe part of Nevada owned by the Bureau of Land Management. And she'd be content with an M 4.0 or so (though some of her peers are pushing her to try for an M 6.0).

She hasn't petitioned the government for permission yet. But she'd like to. She hopes you don't mind.

TWO

A WEALTH OF WATER
AND ROCK

DAM BUSTING WITH
SENSURROUND G-FORCES!

We may have bad memories of them, but Americans love natural disasters—at least on the silver screen. That's been true where earthquakes are concerned for a long time. Take *A View to a Kill*, Roger Moore's last gig as James Bond. In it 007 goes up against a psychopathic genius hell-bent on destroying Silicon Valley. His dastardly plan? To set off a series of earthquakes on the Hayward Fault, thereby flooding all the microchip manufacturers in the area and leaving him with a monopoly. His plan would have worked, had Grace Jones not come to the rescue—at least, that's what we're led to believe by the film's script.

Roger Moore received, at best, mixed reviews for his performance in the film (even he admitted that, at fifty-five, he was far too old to play Bond, an assertion with which Sean Connery gleefully agreed). But the film itself has been celebrated by natural and applied scientists as being one of the "most sciency" in the 007 oeuvre. For years, Steve Dutch, now professor emeritus at the University of Wisconsin–Green Bay, has run the webpage "The Science of James

Bond." He says he gives the film credit for coming up with the conceit of the earthquakes, which he finds more plausible than magnetic fields or seismic wave generators, both of which have been used by other cinematic criminal masterminds and at least one ill-conceived superhero: Gin Genie, who can direct seismic waves at her enemies at magnitudes in direct proportion to her blood alcohol level, leading one pop culture observer to deem her "a superhero that's so crappy she's basically a villain anyway."

Plenty of other films have made big ripples just by depicting natural quakes. Take *San Francisco*, a musical melodrama in which Clark Gable plays the rapscallion Blackie Norton, who survives the full impact of a brick wall collapse during the 1906 quake to find the love of his life, an opera singer who pacifies fellow quake survivors by singing arias while wearing a spotless satin ball gown. Almost forty years later, Hollywood released *Earthquake,* a 1974 disaster drama starring a young Charlton Heston as Stewart Graff, a philandering construction engineer who saves countless high-rise office employees before being swept away in the sewers of Los Angeles after the catastrophic failure of the Mulholland Dam.

The film's producer wanted it to be more than just a movie: He wanted an "event." And so he developed a sound system he dubbed Sensurround: basically, a huge bank of speakers dialed to a low frequency that would shake the theater. Advertisements for the movie included a warning:

> *ATTENTION! This motion picture will be shown in the startling new multi-dimension of Sensurround. Please be aware that you will feel as well as see and hear realistic effects such as might be experienced in an actual earthquake. The management assumes no responsibility for the physical or emotional reactions of the individual viewer.*

He supposedly entertained the idea of also dropping big chunks of stiff stunt foam on moviegoers, but eventually bagged that idea—perhaps after a quick conversation with his legal team. As it turns out, the sound system was more than sufficient. Hollywood film writer David Konow reported that a group of Nicaraguan refugees who had survived a recent earthquake were watching another film in a theater next to the one showing an early screening of *Earthquake*. When the Sensurround kicked in, they leapt out of their seats and ran from the building, thinking they were experiencing another life-threatening tremor.

Most moviegoers ate up the experience, however, even though the film offered an apocalyptic ending with little hope for redemption.

But perhaps the best known of all San Andreas depictions is the eponymously named 2015 film featuring Dwayne Johnson and Paul Giamatti. It grossed $53.215 million its opening weekend alone.

In one of the film's opening scenes, Giamatti's character, Dr. Lawrence Hayes, and his faithful scientific sidekick, Dr. Kim Park, attempt to prove that they can predict earthquakes by monitoring magnetic pulses before a quake. The fictional duo set up their observation station at the Hoover Dam, where rumblings from a previously unknown fault have been going on for days. Park collects data from inside one of the dam's air shafts, while Hayes gleefully records it from the tourist-packed top of the dam. They confirm their prediction theory just as a magnitude 7.1 earthquake crumbles first the Hoover Dam Bridge and then the dam itself, sending billions of tons of water down the Colorado River and killing Park in one of Hollywood's best depictions of impalement by exposed rebar.

If you really want to try the patience of Hoover Dam tour guides, ask them about the plausibility of that scene. I did while crouched in the same ventilation shaft where the fictional Park supposedly took his readings. My guide for the day was Zane, an endlessly cheerful retiree who got bored after twenty-five years of retail-management experience and is now one of the reigning experts on all things

Hoover Dam. Touring the facility with him is a lot like spending the day with a fitter version of John Candy. When a member of our group asks how many employees work at the dam, he first quips, "Half of them. It's a federal facility, after all." When pushed for a straight answer by the same guest, he says he can't answer for security reasons, but "Google it and you'll find something pretty close to the right answer" (I did and came up with 250, forty of whom are currently eligible for retirement).

Our tour takes us through the dam's power plant over 500 feet down inside Black Canyon. We slither through tunnels and air vents and repeatedly pack ourselves shoulder to shoulder in an elevator that looks like it was designed for far fewer people. "Trust me," Zane jokes whenever someone looks worried about something. "I work for the government." And then he laughs—for a really long time.

Zane's countenance changes considerably when I bring up the subject of *San Andreas* and the destruction of the dam.

"Totally far-fetched," he says. "Absolutely none of it was believable."

Rose Davis, the Bureau of Reclamation's spokesperson, took that outrage one step further when the film was released: She told newspaper reporters she was going to boycott the movie outright. She also made it very clear that the scene wasn't even filmed at Hoover Dam. The Bureau of Reclamation insists on full script approval before allowing any movie to be shot on location. It summarily rejected the script for *San Andreas*.

"We didn't want to participate in that kind of far-fetched thinking," Davis told the *Las Vegas Review-Journal*.

During production, the cast and crew of *San Andreas* weren't even allowed to visit the dam. Zane thinks they probably came anyway and posed as tourists, but neither he nor I could confirm that.

What I can confirm is that the Bureau of Reclamation had no problem allowing *Transformers* to film on location at the dam. In that flick, two warring races of gigantic intergalactic robots descend to earth after their planet is destroyed. The body of Megatron, the malevolent supreme leader of the Decepticons, is discovered in an Antarctic crevasse just as the Great Depression is about to take hold. The US government responds by building the Hoover Dam to hide the thirty-five-foot robot and his evil energy field. The dam spends its next eight decades overseen by a governmental agency known as Sector Seven, dedicated to extraterrestrial threats so secret not even the secretary of defense has heard about them—or the agency itself.

"It's, frankly, a more plausible scenario," says Zane.

Nevertheless, the prospect of *any* earthquake on San Andreas is enough to get seismologists, physicists, and emergency managers thinking about all kinds of doomsday scenarios. In 2008, a team of them conducted an extensive study of what would occur in the event of a magnitude 7.8 earthquake on the southern section of the San Andreas Fault Zone. Lucy Jones, former seismologist for the USGS and earthquake-science superstar, was the lead author on the study, which began not long after Hurricane Katrina devastated the Gulf Coast. She says the aftermath of that storm was very much on the team's minds as they began their work. "In Katrina," she later wrote in the completed study, "we saw how catastrophe can strain the fabric of society and lead to decades of economic disruption. Since Katrina, we have distinguished between a natural disaster—an inevitable event such as a hurricane, flood, wildfire, or earthquake—and a catastrophe, which occurs when a disaster disrupts a large region and the effects continue for decades."

Part of what would be so catastrophic about a quake hitting the southern San Andreas Fault Zone is its effect on structures like dams, even those built to high safety standards. And dam failures, says Jones, have historically been some of the deadliest and most expensive results of a quake. The authors of the 2008 ShakeOut study estimate that about thirty dams within a fifteen-mile radius of San Andreas could "experience damage serious enough to cause safety concerns."

Is it really so impossible to believe that the Hoover Dam could see a similar fate?

The Hoover Dam was originally slated to be located twenty miles upriver, in Boulder Canyon. When surveyors found evidence of geological faulting, they moved it downriver to a second site. But that site proved too steep to move around construction materials and workers efficiently, and so it, too, was rejected. Eventually, they settled on Black Canyon.

While engineers were finalizing their plans, Los Angeles witnessed one of the country's worst dam disasters. In 1928, the St. Francis Dam—itself an engineering marvel and LA's first massive concrete impoundment—failed catastrophically, killing at least 450 people and making it second only to the 1906 San Francisco earthquake in the echelon of California's deadliest disasters. The dam's creator, Los Angeles Department of Water and Power's chief engineer, William Mulholland, would swear for the rest of his life that the dam failed only because of saboteurs or unexpected seismic activity. Inquest teams and juries disagreed: They claimed the dam failed because it was bolted to weak rock, which included the San Francisquito Fault.

Public outcry over the dam's failure prompted the Bureau of Reclamation to establish an independent panel of experts to review their plans for what would eventually become known as the Hoover Dam. At the time, engineering considerations were based almost entirely on mathematical probability. Seismic analysis of the dam was conducted primarily by H. M. Westergaard, a professor at the University of Illinois, who devised a series of equations to determine how the dam would respond to ground motion and a change in water pressure. His conclusion? The dam would shake and maybe even sway, but it would hold.

FDR dedicated the dam in September 1935. At the time, the Bureau of Reclamation was so confident in the dam's integrity that they invested heavily in an art deco decorative motif in even the deepest regions of the structure. To this day, places like the generator room, located in the belly of the dam, are floored in hand-cut porcelain tile with brass inlays. Throughout the structure, copper elevator doors, marble walls, and punched aluminum ceilings greet guests with an aura of extravagant elegance.

"They knew you were coming," Zane tells us as we stride from space to space, the click of our shoes on polished mosaic. "They even filled in floor drains so that women's high heels wouldn't get caught."

What they didn't plan on was the full strain created by over 40 billion tons of water pushing against even the strongest concrete structure. In June of 1937, as the reservoir began to fill, cracks in the dam began hemorrhaging water—over 200 gallons per minute, according to Michael Hiltzik, author of *Colossus: The Turbulent, Thrilling Saga of the Building of Hoover Dam*. Bucket brigades were no match for the surge, so engineers cut huge ditches in the newly poured floors to direct the water out of the dam. That same water cracked any weak areas in the poured concrete and the river canyon itself, exposing previously unknown faults that now wept with runoff. Engineers began to worry aloud about the longevity of the structure. The Bureau of Reclamation responded by initiating an emergency coring of the bedrock below the dam. "The results were shocking," writes Hiltzik. "The bedrock was riddled with faults, shears, and soft rock veins through which water flowed freely. It was now evident that the geological mapping of the canyon in the 1920s and 1930s, viewed at the time as exceptionally thorough, in fact had been woefully superficial."

But that was just the start. As water continued to pour into the dam, surveyors found insufficient support: The dam's grout curtain, a row of 150 boreholes sealed with pressurized grout and intended to protect the dam's foundation from water inundation, was failing. It would take an emergency allocation of $2 million (over $33 million today) and nearly a decade of constant work to shore up the weaknesses and avert disaster. Even with this hefty price tag and the work it required, the Bureau of Reclamation managed to keep the breach and subsequent repair almost completely out of the public eye: At the time, only a small group of engineers and construction workers even knew it was happening. And you won't hear anything about it at all on the tour.

You also won't hear about the quakes the reservoir itself set off in the region. Water is exceedingly heavy. And when there's enough of it, it can have the same force on the earth that tectonic activity can. As Lake Mead, the reservoir created by Hoover Dam, neared its capacity in the late 1930s, the pressure of over 40 billion tons of water began to deform the bottom of the reservoir, shifting the strain on canyon walls and the surrounding area. The first resulting earthquake occurred in 1936. It was soon followed by another. And another. By the time the Seismological Society of America convened for its annual meeting in 1938, the landscape around the Hoover Dam was showing sustained seismic activity. The SSA petitioned the Bureau of Reclamation for permission to install a network of seismographs around the dam. By then, the bureau was more than happy to oblige. The network arrived just in time. On Thursday, May 4, 1939, the most powerful quake yet struck the region. Measuring M 5.0, it was felt as far away as Los Angeles and Mexico. Aftershocks rippled for days. A month later, a second 5.0 quake rattled the region. Geologists in the US Coast and Geodetic Survey scanned any and all public records they could for the region, and couldn't turn up a single quake prior to the installation of the dam. The causality seemed clear.

For the next thirty years, changes in water levels in the new reservoir continued to set off earthquakes, some of which came in the form of swarms rivaling those in Yellowstone. The USGS estimates that, in the first forty years after its completion, Hoover Dam witnessed over 10,000 earthquakes, no small number of which were strong enough to rattle Las Vegas (that number, in all fairness, slacked off considerably after the study was published in 1976).

Hoover Dam isn't alone in this phenomenon. When conducting a geological audit for a power plant, the Nuclear Regulatory Commission found thirty earthquakes in the South Carolina region that were either reservoir-induced or "probably reservoir-induced." Major earthquakes have also occurred at reservoirs around the world, and several of them have been far more powerful than the strongest ones witnessed in southeast Nevada. A magnitude 6.3 quake occurred during the filling of a Greek reservoir in 1966. A year later, an M 6.5 quake occurred under very similar circumstances in India. It killed nearly 200 people, most of them women and children, and injured 1500 others. The dam also happened to be part of the largest hydroelectric power plant in India, so its failure knocked out power in huge swaths of the country, including in Bombay. The dam itself developed some cracks, but they were grouted without much problem. Since then, the dam has continued to witness earthquakes; the latest one of note was an M 5.1 in 2009.

A similar reservoir in Tajikistan is particularly interesting to seismologists, since it seems that changing the water height by just a few meters there is enough to trigger quakes. And in 1963, a massive rock slide at the Vajont Dam in Northern Italy created a tsunami-like wave over 300 feet taller than the crest of the dam. Residents in the area said that smaller slides had become common since the construction of the dam, but no one was prepared for the 350 million cubic yards that rained down there that October night—about five times the volume of debris that created Hebgen Lake.

So far, Hoover has weathered the quakes just fine. You'll find plenty of cracks inside, but that doesn't bother anyone there much. "There are two types of concrete," Zane tells our tour group. "The kind that has cracks in it and the kind that soon will." He comforts us with the observation that not all cracks are created equally. The ones in your coffee cup will probably spring a leak. The ones in the dam probably won't: They're too shallow, and the pressure of the concrete around them is too great.

Nathaniel Gee, supervisory civil engineer at the Bureau of Reclamation's Dam Safety Office, is responsible for keeping Hoover Dam safe. I sit down with Gee and Rose Davis, Bureau of Reclamation spokesperson, to find out how he does so. It's been over a year since the release of *San Andreas*. Davis says she still hasn't seen it, but she is thinking she might consider it, now that the film was available on premium cable.

"Disaster movies just aren't my thing," she explains. "Especially dam disasters."

I tell her that, much to the delight of the adolescent males in my life, I'd watched *San Andreas* and *Transformers* back-to-back in preparation for our meeting. I suggest she might want to do it the same way—so long as she has some good snacks on hand.

"And a cocktail," she adds.

I can't argue with that.

Gee, 33, just sits quietly, with a pleasant smile on his face. He doesn't drink, and he'd rather write historical romantic satire in his spare time. But he does offer that he and his family live in a house built by one of the workers who arrived in Boulder City by the thousands to construct the dam back in 1931. Gee sometimes carries with him a letter written by that worker to his sweetheart back home, complaining that the dam-building crews were shorthanded and that quitting alcohol was making him awful sick.

When Gee is on the job, he works about as hard as his home's original occupant. Gee is responsible for ensuring the integrity of the three dams on the lower Colorado River, including Hoover. Sometimes, he does this by rappelling down their faces (that's 726 feet in the case of Hoover). Other times, he scuba dives to their base.

Our meeting takes place just after the fortieth anniversary of the Teton Dam failure, the worst in Bureau of Reclamation history. Construction on the Teton Dam, believed to be the most advanced in dam technology of the time, concluded in November of 1975. The bureau began filling the reservoir behind it that same month. Six months later, it was near capacity. On the morning of June 5, 1976, workers began noticing small leaks in the base of the dam. The number of leaks increased steadily over the ensuing hours as workers struggled to plug them. Nothing worked. By 11:00 A.M., sheriffs in the surrounding communities had received an ominous warning: The Teton Dam was going to fail. Less than an hour later, 80 billion gallons of water burst through an enormous hole in the dam, creating a massive surge that killed at least eleven people and thousands of cattle.

The largest earthquake recorded in the area during construction of the dam was M 2.2—not even strong enough to be felt by most people. There was no record of any seismic activity in the days leading up to the quake, nor was there any significant rainfall or other source of flooding. Instead, says Gee, the dam suffered from what people in his line of work call "sunny-day failure"—an event caused by a structural flaw in the dam rather than a precipitating event.

There had been seismic near misses before. Take the 1964 Alaska quake. At the time, the Eklutna Dam power plant was operating near capacity and pumping out electricity for much of the region. The earthquake itself occurred on March 27, around 5:00 P.M. local time. Its rupture damaged the transformers and circuit breakers connected to the dam. Before the shaking was over, much of the region had been plunged into darkness. Supervisors shut down the plant immediately for inspection. Within a half hour, it was back up and operating, and electricity was transmitting from the undamaged circuits. But early the next day, water levels in the penstocks (the pipes that deliver water from a reservoir to the power-generating turbines) began to fluctuate.

Initially the problem appeared merely to be caused by debris created by the quake—everything from branches to parts of houses and vehicles that were clogging up the intake—but it was hard to tell for sure.

An Alaskan March is still a long way away from spring thaw. The area around the dam was packed with snow and ice. It wasn't until May that a full inspection of the dam could be completed. In the meantime, the hydroelectric plant kept running at full capacity—it was the only one in the region operable after the quake. Once the plant was shut down for its inspection, crews were shocked to discover that the dam's intake structure had moved more than forty inches during the quake and that rocks as big as basketballs were clogging the penstocks. The dam itself had "settled" to a dangerous degree. The dam was declared unsafe and the reservoir was drained. Eventually, a new dam was built downstream.

In 1971, an M 6.7 earthquake rocked the San Fernando Valley, including the Lower Van Norman Reservoir, a major linchpin in the Los Angeles viaduct and drinking water system. The quake shook the region for less than a minute, but that was more than enough time to damage the reservoir's embankments, causing over 1000 feet of the retainers to crumble into the reservoir. The shaking also caused the top thirty feet of the dam—over 800,000 cubic yards of material—to shear off as well. Water lapped at the top of what remained of the damaged structure, and each subsequent aftershock caused more of the dam to crumble. At the time of the quake, the reservoir was only half full. Had the reservoir been nearer capacity, the force of the water would have been more than enough to burst through what remained of the dam. Even at the reduced height, the reservoir posed enough of a threat to the communities living below it that the city made the decision to evacuate the 80,000 people living in its flood zone. For days, engineers tried to further reduce the reservoir levels by cutting holes in the dam and surrounding aqueducts, but they succeeded in reducing the overall levels by only a few feet.

Those near misses did little to change how the Bureau of Reclamation approached dam safety. Prior to the Teton disaster, early modeling was primitive at best. When it came to predicting the integrity of Hoover Dam, says Gee, design engineers basically built a rubber block intended to resemble the dam and then ran a series of calculus equations to determine what it could withstand. They then put that block on a table and repeatedly knocked it over to see if their equations were accurate. So far as he and I can determine, no one tested to see what would happen in the case of an actual earthquake.

After the failure of the Teton Dam, Congress passed the Reclamation Safety of Dams Act, which ultimately created Gee's job, along with the bureau's entire dam safety program. A few years ago, the bureau produced a video explaining how this tragedy led to better safety standards. Over a soundtrack of jaunty electronic music and reels of archival video, the narrator admits that, prior to Teton, the bureau had developed "a pattern of insular decision making and centralized processes" in which design engineers and construction heads never consulted with one another. If one branch wanted to talk to the other, they first had to fill out a request form and fax it to management. If they wanted to make a long-distance call, that required another form.

The failure of the Teton Dam changed all of that. Now each project includes a principal design engineer and a principal geologist who work in tandem with each other. Plans must undergo peer review, and every dam must also include a plan for both risk management and emergency action. But perhaps the biggest change was the implementation of a mandatory comprehensive facility review of each dam on four- and eight-year timetables. One of the express purposes of that review cycle is to assess seismic loading of a dam. Why? Because, says Jim Mumford, retired coordinator of dam safety at the bureau, "there are very few structures that we deal with as a society that could have a greater impact to a community than the dam. It's a major responsibility nobody takes lightly."

That's particularly true where seismicity is concerned. A Bureau of Reclamation technical paper noted the irony of our lack of data concerning seismic dam failure in bureau dams. None has failed catastrophically in an earthquake. "This is comforting in itself," wrote the paper's authors, "but also means that seismic failure modes for concrete dams are not well understood."

〰

Keeping track of an aging dam is no easy enterprise. Hoover is now seventy-five years old—and, says Gee, that means most of its auxiliary workings (things like pipes and valves) are well beyond their design life. Further complicating the situation is the fact that Gee, like seismologists, can't actually see a lot of what he's supposed to be monitoring. The technical term for this is "inaccessible feature of use." What it means is that the guts of the dam are encased in concrete or under 500 feet of water or, in some cases, so encrusted with marine life (invasive mussels are a particular problem, says Gee) that they are hard—and sometimes impossible—to inspect. And so instead he works with a team of engineers and seismologists at the Bureau of Reclamation's Denver office. Their job?

"Millions and millions of computations looking at millions and millions of variables," says Gee.

Even without them, says Gee, Hoover would be the least of his worries. It's what engineers call an "arch dam"—one that uses the arc of its design to distribute the immense load of water. Arch dams tend to be pretty thin, even at their base. It's not uncommon to find a huge one with a base as narrow as thirty or forty feet. The base of Hoover is 600 feet thick.

"They really overdesigned it," says Gee. "Had they built Hoover twenty years later, they would have had a lot more confidence in the design. But at the time, it was pretty new technology. So they built it as conservatively as they could. They wanted to make sure it was going to stick around."

How overdesigned, I ask?

"It's invincible," he says. At that point, Rose Davis cuts in.

"I don't think we want to say 'invincible,'" she advises.

"Okay, not totally invincible," he concedes.

I ask what it would take to bring down the dam. They both laugh good-naturedly.

"We're never going to answer that question," says Davis. "It'd be like giving a terrorist a blueprint for destruction."

That's fair, I concede.

What about earthquakes?

As it turns out, concrete dams like Hoover perform better than expected when confronted with a quake. There are a couple of reasons why. By their very nature, they exist to withstand pressure and force. Redundancy is added to allow for this. That means that, even if an area of the dam is compromised, there's enough volume of concrete around that area to compensate. And because it's already wedged into a canyon, it has added stability.

Contemporary concrete dams that have experienced significant earthquakes have shown very little sign of leakage. Cracking that has occurred tended to be near the top of a dam, and since very few reservoirs are kept at capacity, those cracks are often above the waterline.

If there is going to be a problem, says Gee, it's going to be on the auxiliary parts of the dam system—the intake towers that suck water into the penstock pipes, the penstocks themselves, the spillway tunnels. Those are the parts of a dam not made of 6 million tons of concrete.

"I'm pretty certain no earthquake can bring down Hoover Dam," concludes Gee.

I must seem a little disappointed by this revelation, because Gee next offers me a consolation prize.

"We did have to do some seismic remediation at Hoover," he offers. "In 2007, we had to retrofit the Snacketeria."

―――

In general, what we do know about dams and seismicity suggests that, even globally, these structures perform really well in earthquakes. During the Hebgen Lake rupture, that dam was overtopped again and again. Still, it held. That's often the case. The US Society on Dams is a consortium of academics, government officials, and industry insiders. They estimate that, worldwide, only about a dozen dams have ever been destroyed by earthquakes; most of them were old and poorly constructed out of natural soil. About a dozen concrete and embankment dams have been severely damaged during earthquakes—enough so that they needed to be closed or replaced. But when you figure that there are over 80,000 dams in the United States alone, those numbers start to look pretty good.

To understand why, it's best to go back to your high school physics class or the last time you tried unsuccessfully to pry yourself away from binge-watching a new TV series. In both instances, you had to confront the concept of inertia—a body's tendency to want to stay either at rest or in motion, depending upon the state in which you found it. Large structures, just like most Americans, are really good at rest. Earthquakes disrupt that state. To figure out just how much they disrupt, you need to know an object's mass and then multiply that number by its rate of acceleration. Acceleration is often measured in terms of gravitational force—or g, for short. A navy fighter pilot taking off from an aircraft carrier experiences about 4 gs. A race-car driver going around a tight turn will experience a little over 1 g—the same as your average bungee jumper.

Earthquake shaking can also be recorded based on g-force. It's the job of the dam's design engineers to figure out how much of that force the structure can withstand. Just like a bungee jumper, structures like dams (and houses and bridges) experience different types of motion. Of particular interest to engineers is what's called "peak ground acceleration" (PGA). The greater the peak ground acceleration, the higher the likelihood that a structure will be damaged. It gets complicated when you start looking at things that are tens of stories tall or more: They often have a different response to PGA at the top than they do at the bottom. Also important to engineers is the range of motion. A structure doesn't move in just one direction during ground shaking: It can get torqued or twisted in multiple ways and can also be raised or lowered. Engineers also need to know what the maximum shaking might be for a particular area, and how likely that shaking is to occur. Usually, this is based on a specific return rate: say, the highest possible earthquake in 100 years or in 10,000 years (which is the standard used by the Bureau of Reclamation).

A lot of their baseline information comes from the US Geological Survey's seismic hazard maps. For decades, the USGS has forecasted future earthquake risk by way of these illustrations. Updated every six years, these maps provide an account of the likelihood that a certain area will experience a certain level of shaking in fifty years. The "hazard" they illustrate isn't just the magnitude and location of potential quakes, but also how often those quakes have historically occurred and the kind of rock and soil the earthquake waves would pass through. This is an important point. As laypeople, we tend to talk about earthquakes almost entirely in terms of their magnitude. But geologists and engineers and emergency planners are at least as interested in their intensity. Intensity takes magnitude into account, but also the distance from a given earthquake (the closer you are, the more you feel the shaking) and the type of ground beneath your feet (remember, hard rock like what is found east of the Rockies is much more effective at transmitting waves than what you might find in Arizona; soil prone to liquefaction is going to do a lot more damage to a structure than other types of ground). Intensity is usually talked about in terms of the Modified Mercalli Intensity Scale, a fabulously descriptive rubric any former liberal arts student can get their head around. Level III, for instance, is described as "vibrations similar to the passing of a truck." Level VI? "Felt by all, many frightened. Some heavy furniture moved; a few instances of fallen plaster. Damage slight." And, finally, Level XII: "Damage total. Lines of sight and level are distorted. Objects thrown into the air."

To create the hazard maps, the USGS begins with what are called "historical maximum seismic intensities." They then overlay the rate at which these quakes have occurred. They also take into account what is known about existing faults in the area. What results is a map that looks a lot like one you've probably seen displaying the predicted high and low temperatures for a region or the nation: circles of red and orange over hot spots, bands of yellows and blues and greens where things are increasingly cooler. These maps get used for everything from dam design to the insurance rates for your house to whether FEMA is going to invest in public service announcements that will play as commercials during your favorite game.

Art Frankel has been at the center of that project for thirty years. He works out of a tiny office at the University of Washington that could be considered, at best, spartan. There are few if any personal touches to the space, save for a bottle of orange Gatorade on his desk. He dresses in the uniform of the geologist: jeans, button-down cotton shirt, inexpensive watch, USGS fleece vest. Frankel is shy. But when he starts to talk about forecasting seismic hazard, he gets excited in an understated way. If he tells you that something is "really cool," you know you've reached that threshold.

The USGS began forecasting seismic hazard in 1976, but the

program was radically overhauled in 1993 and Frankel was appointed to lead the program. Paleoseismology was really taking off then, and trenching was providing new information about past events. Today, that kind of study is coupled with computer simulations and 3-D monitoring. Nevertheless, says Frankel, forecasting the potential hazard for a quake is hard. "There's so much that we don't know," says Frankel. "We're dealing with very, very large uncertainties."

One of the biggest problems, he says, is the simple fact that we don't know where most faults lie.

Take the case of the 1967 earthquake in southwestern India. The M 6.5 event left a series of deep cracks in the Koyna Dam, a massive concrete structure standing over 330 feet high and extending over half a mile in length. Soon after the quake, water from the abutting reservoir began seeping through these cracks. Further inspection revealed that various sections of the dam had moved during the quake. The damage was significant—and hugely costly to repair. But what most concerned officials was the fact that when the Koyna Dam was built, the area was believed to have no seismic activity. As a result, the dam was built to only minimal seismic standards.

That's not an uncommon situation, says Frankel. One reason is that we are finding new faults all the time. In fact, he says, we've barely scraped the tip of the iceberg when it comes to knowing just how many there are. And then there's the problem of knowing which ones are loaded or can be reactivated. That's particularly true east of the Rocky Mountains. You might remember from the introduction to this book that the USGS quaternary fault database has records for 2100 faults that have ruptured the surface in the past one million years (a lot of this information was gleaned through the efforts of paleoseismologists). Frankel says we just can't know how many of these faults are currently active. And he doesn't know what percentage of the total faults in the United States that this number comprises, but he's "sure this is a very small portion of the total number of faults." Just how small a portion? He says it's impossible to say for sure.

In 2015, a series of earthquakes struck across northern Idaho and Montana. A resident USGS geologist was interviewed for the local paper and asked to explain what had happened. "I don't think we understand what fault this earthquake occurred on," he said. "There's a USGS map of active faults across the US and it doesn't have anything for northern Idaho. That doesn't mean there's no seismic activity, we just don't have a ready fault to pin this on."

That's part of why hazard seismologists like Frankel depend so much upon the research being done by people like Judi Chester and Charles Merguerian to advance these hazard maps. Every time a new fault is discovered, it gets added to the map. Every time there's an earthquake, it gets added to the map. Neither of those two things happens very often. But geologists and their very accommodating graduate students nevertheless go out into the field every summer trying to make more sense of the faults we know about. That goes into the maps, too. That, says Frankel, is part of the reason why these maps keep getting updated: Every time we learn something new about where faults lie or how ground motion works, our best guess about hazards changes. It's a pretty big deal when the new ones get revealed. That last happened in 2014. Those models used about twice as much ground shaking data as the previous ones.

With the release of the most recent analysis, the forecasted likelihood of magnitude 6.7 and higher earthquakes in California decreased (though geologists there still say the possibility of such a quake remains at "near certainty"). Meanwhile, the likelihood of an 8.0 or greater has increased. Much of the eastern United States was also nudged up in terms of hazard possibility, along with big sections of the Midwest.

In 1998, the Bureau of Reclamation issued a technical memo on the Hoover Dam based, at least in part, on those hazard maps. They based their calculations on the highest possible 100,000-year event (they're really confident in their longevity over there), which they figure is an M 7.3 quake. Gee thinks even that's overstating the case: The USGS puts the likelihood of such an event occurring in the next 100,000 years at less than 2 percent; they figure there's maybe a 30 percent chance it could happen in the next *million* years. Gee says he doesn't think even that million-year event is going to do much damage to Hoover Dam.

How do they know?

Design engineers anticipate earthquakes using two different methods. The first is deterministic. It takes a specific magnitude earthquake and plots its effects on a local area. In the case of Hoover Dam, they'd take that M 7.3 number and then figure out exactly what it would do to the dam. It doesn't matter that the likelihood of that quake in our lifetime—or the dam's lifetime—is really low. It's ultimately a game of "What if?" What if that quake were to occur? How would the dam do? A complementary approach is probabilistic assessment. Much like the USGS hazard maps, it takes into account everything known about historic seismicity, local geology, and fault systems to determine the likelihood that an event will occur during a particular interval.

Still, there are surprises. In 2010, on Easter Sunday (quakes seem to love Easter Sunday for some reason), an M 7.2 earthquake ruptured in northern Baja California. It was a shallow one—about six miles deep. The intensity was enough to crack part of the bypass channel on the All-American Canal, which provides water for irrigation to a swath of Southern California and Mexico. Repairs took over three years to complete.

That, says Gee, is what we're likely to see in similar quakes. The Bureau of Reclamation recently concluded a seismic audit of its dams on the Colorado River. They found several of their dams to be "unsafe by modern engineering standards," which is to say that a dam could be damaged were the area to witness its maximum credible earthquake. In some cases, they added new spillways and flood-control space; in other cases, they required the total height of a dam to be raised to accommodate sloshing. In cases where soil liquefaction is a concern, they're excavating underneath dams and replacing the problem soil with concrete and more stable fill.

Of particular concern to the bureau has been Bartlett Dam, which sits less than fifty miles northeast of Phoenix. Were it to fail, the city would be in serious trouble. Currently, the bureau is in the process of reinforcing key features of the dam and raising it by nearly twenty feet. They're also creating auxiliary spillways that would help move water past the dam in the event of a major earthquake.

That's true for many of the hydroelectric dams in the Pacific Northwest as well. They're probably strong enough to resist an earthquake, but their auxiliary workings and power transmission lines may not be. And that could still spell big headaches for the people who rely upon both. But the biggest dam headache of all has been, perhaps not surprisingly, in quake-prone California.

Auburn Dam is the most famous dam you've never heard of. And I can guarantee you've never visited it, either. That's because it never existed. But it almost did. And that alone was enough to set off one of the biggest controversies in the history of American water rights. At the center of it all? An earthquake.

When it was first proposed in the late 1960s, the dam was scheduled to be one of the largest in the country—rivaling even Hoover in terms of its size and scope. The Bureau of Reclamation chose a wide canyon on the American River in California's Central Valley as the location for this behemoth and the massive reservoir it would create. There, the dam would capture spring runoff from high in the Sierras and hold that water for agricultural purposes. As an initial outlay, the federal government spent about $300 million testing the area, completing designs, and laying the framework for construction. But in their preliminary work, they found the rock around the dam site unstable and had to relocate the original tunnel intended to temporarily reroute waters for construction.

Initial plans called for the dam to be an earth-fill design requiring 25 million cubic yards of dirt and rock that would be skimmed off the surrounding area. That's a lot of dirt. By some estimations, it would have left huge swaths of land naked to the bedrock, including areas around the cities of Auburn and Sacramento. That proved unpopular with local residents, who thought their landscape should stay where it was and not be used to back up a local river, even if it was one that had a propensity for flooding and sometimes being a nuisance.

So the Bureau of Reclamation went back to the drawing board and came up with a design for a concrete dam. When finished, the Auburn Dam promised to be the longest of its type in the world. Construction began in the spring of 1972 and was instantly met with tragedy, when unstable rock in a tunnel collapsed and killed one of the workers there. Still, construction continued for another three years.

And then an M 5.7 earthquake ruptured less than fifty miles north of the construction site. The Bureau of Reclamation had done a seismic study before they began work on the Auburn Dam, but they hadn't taken into account any active faults in the region. Residents in the area grew nervous. So did the state of California. Were this new dam to fail, it would send a massive wave of water that would take out the Folsom Dam downstream and quite possibly threaten the entire city of Sacramento. The Bureau of Reclamation offered to authorize additional seismic studies, certain they would prove that the dam's design and location were sound. But that's not what happened. Instead, a geological subcontracting firm hired by the bureau found huge rifts near and around the epicenter of the recent quake and additional active faults dangerously close to the dam, including one that was just three miles away. At least one of those faults, they said, could certainly prompt an earthquake with a magnitude of 6.5. That'd be more than enough to slide the dam's foundation by nearly a foot. They also found the likelihood of reservoir-induced seismicity at the dam to be anywhere from 2 to 30 percent. So the Bureau of Reclamation decided to hire a different contractor to see if they could get a different conclusion. Meanwhile, both the USGS and the state of California had launched their own investigations into the dam. The USGS also thought the initial contractor was mistaken, but not in the way the Bureau of Reclamation had hoped. They actually thought a much larger quake— one with a magnitude of 7.0—was entirely possible. That, they said, could push the foundation of the dam more than three feet and easily spell catastrophe.

The Bureau of Reclamation responded to this news by reasserting their alliance with the original contractors. A 6.5 quake wasn't so bad after all, and neither was a foot of distortion. But that wasn't good enough for the state of California. They were sympathetic to a point, but not to a foot.

Amid all this dithering, then president Jimmy Carter decided he'd had enough. The dam was already a money sponge, and he'd yet to see any evidence it'd pay for itself, even if it were seismically sound. He cut funding for the project. Water-rights supporters in the West grew vociferous in their dissent: Clearly, Carter hated dams. And agriculture. By then, Carter was a lame-duck president. He told the bureau that if they could build the dam safely, he'd let it happen. Let Reagan deal with the rest of the mess.

And so he did—for all eight years of his presidency. By the time he took office, the Auburn Dam came with a price tag of $3 billion: exponentially more than initial estimates. Were it to be built, it would become the nation's most expensive dam by far. Reagan agreed. So did Bush. And Clinton. And the next Bush. They all let the project languish largely forgotten, though every once in a while a California senator or congressperson would raise the issue again, mostly as part of one campaign or another. After Hurricane Katrina, it looked like the dam would get some real traction again as state officials worried the scene in New Orleans might be replicated in Sacramento, but those nagging seismic questions kept upping the price tag of doing anything. By 2004, building the dam to meet those concerns now came to a grand total of anywhere between $6 and $10 billion (even adjusted for inflation, the total price of Hoover was just $777 million). Finally, in 2008, California revoked rights to the American River from the Bureau of Reclamation. They'd had enough. And so had the nation.

Water is a contentious issue in the West. Quite possibly the most contentious issue, when you consider debates like those surrounding Hetch Hetchy and the Glen Canyon Dam. There is no middle ground: You either love dams or hate them. And people on either side are often ready to do just about anything in the name of their cause, particularly if it means maligning the other side. That's the wild thing about the Auburn Dam. For what was quite possibly the one and only time in the history of the American West, it brought those two sides together. Ask someone from either side, and they'll tell you it was, to borrow a phrase from the Bureau of Reclamation, "an unmitigated disaster." All because of one moderate earthquake.

━━*ww*━━

As the nation's infrastructure continues to age, the potential hazard created by dams increases. The average age of a dam in the United States is around fifty-two years. The American Society of Civil Engineers currently estimates that, of America's some 84,000 dams, at least 4000 dams in the United States are "deficient." Half of those deficient dams are also rated "high hazard" using the FEMA designation. That number will continue to increase as population centers grow and the dams themselves continue to age.

These dangerous dams, by and large, are not Bureau of Reclamation dams. Most of them are privately owned. And that's a real problem. Some are regulated. Others are not. The Association of State Dam Safety Officials says it will cost about $21 billion to repair all the deficient high-hazard dams that we know about. Regulation of these dams falls to state dam safety officials. Every state is different. Alabama, for instance, doesn't have a dam safety regulation program. South Carolina has one, but has only two safety inspectors for its more than 2300 dams.

Even more troubling is the fact that over a third of the nation's dams do not have an emergency action plan. That includes dams in states with high seismic risks like Oregon, where twenty-four high-hazard dams are without action plans. An investigative project completed by journalists there found that nineteen of those high-hazard dams are in "poor or unsatisfactory" condition.

In California, work is under way to retrofit some of the most seismically vulnerable dams, but that work can come with an unexpected price. Take the Anderson Dam near Silicon Valley. The 240-foot earthen dam was built in 1950. At the time, geologists saw no evidence of active faults in the area. They also felt confident that the dam was secured on solid bedrock. Geologists mapped the area for new development in the 1970s and found evidence of a sizable fault known as Calaveras. In 2011, a study of the dam found that it would not survive a significant quake on the Calaveras Fault. Were it to fail, a thirty-five-foot wave of water would sweep downriver, inundating the town of Morgan Hill in minutes and flooding all of San Jose in just a few hours. Similar findings occurred at the Calaveras Dam, located just 1500 feet from the eponymously named fault. In that case, water authorities drew down the dam to less than 40 percent capacity and made plans to create an enormous earthen replacement farther downriver: a project that required the relocation of over 7 million cubic yards of earth. When it was found that that earth contained naturally occurring asbestos, the project became one of air quality and hazmat monitoring as well.

Even with all of these complications, the Calaveras solution is in many ways easier than the one facing the Anderson. Located in a steep canyon, that structure cannot be re-sited. And so authorities began the arduous work of plotting a retrofit while keeping the reservoir behind it at levels that would not threaten the water supply of the valley. Work was expected to begin in 2016 but was delayed after surveyors found additional faults that run right through what a lead engineer called "the gut of the dam." Meanwhile, what they thought was bedrock underneath the dam turned out to be a dangerous combination of sand and gravel, which is notoriously unsteady in a seismic event.

To abandon the dam project would be to deprive a drought-stricken state of more water that it desperately needs. To proceed could put those who live in the vicinity at risk. Weigh the odds. What would you do?

THE BIG MUDDY

A little before 2:00 P.M. on a sultry summer afternoon, tourists and history buffs begin accumulating in the gift shop of Beale Street Landing, a steel-and-glass edifice at the base of Memphis's most iconic strip. They, which is to say *we*, are a drowsy bunch, bellies full of corn bread or taffy, slowed down by the gravitational pull of the region's infamous humidity. But the gift shop's air-conditioning is running on overdrive, and the Elvis tchotchkes keep us placated until a disembodied voice announces it is time to board the *Island Queen*, a hundred-foot paddle wheeler.

Yes, Memphis has the blues. And, yes, you can see marching ducks at the Peabody Hotel. And of course there's Elvis and barbecue that will make you swoon. But if you want to understand what built—and what continues to drive—this area, you have to get up close and personal with the Mississippi. The trim German couple standing behind me as we board the boat can't contain their excitement. Over a mile wide here, the Mississippi is just so huge. And the kitsch of this boat, with its faux Victorian styling, is just, well, so *American*, they tell me.

I agree. When it comes to natural icons, this is one of the country's greatest. Every grade school kid I've ever known loves to rattle off the spelling of this massive river. Grown-ups wax wistfully about floating down its corridor with Huck and Tom. The river itself is visible from space. Our version of the Nile: Heck, even Memphis is named after the home of that great river.

Like the Nile, the Mississippi River begins as a trickle: more of a rivulet than a river, a seep from northern Minnesota's Lake Itasca. But as it flows south, it picks up the water of other rivers—some mightier than it. The Illinois. Missouri. Ohio. And so the same river that ran at just 12,000 cubic feet per second (or cfs) through Minneapolis has grown to 679,000 cfs by the time it reaches the Louisiana border. By the time it reaches the Gulf of Mexico, that same river will include drainage for 40 percent of our nation's waterways, making it the third-largest watershed in the world.

Aboard the *Island Queen*, passengers range from dutifully impressed by this information to downright bored. Most of them stand patiently in line at the concession stand or drape themselves over the boat's railings to get a glimpse of the paddle wheel churning up thick brown river water. A smaller but very earnest group of us crowd around plastic patio tables, eager to hear James, our tour guide. James wields a microphone, a lousy sound system, and a cafeteria-size can of corny jokes. They delight the four retired women sharing my table, who brought with them enough bottles of wine to make any lecture amusing. I admit I'm amused, too, and I'm even okay with the fact that he keeps referring to us as "class" and treating us like we belong in an elementary school one at that. James is full of fun facts about this river that human engineering has created. Really, though, I'm waiting for the big payoff: I want him to talk about what seismic forces have wrought on this same body of water—and how quickly those forces could crush even our most robust attempt to control it and the surrounding area.

As far as geologists are concerned, the real tourist attraction around these parts is what lies *underneath* this mighty river. There sits a rift not so different from the midcontinental one. The life of this rift began just over a billion years ago, when several existing continents (either three or four—paleogeologists aren't sure) collided. What would eventually become North America found itself, at the time, locked together with the landmasses that would eventually become known as Australia, Antarctica, and China. The sum total of this mash-up was a supercontinent known as Rodinia. And the force of the collision was so traumatic that it created Himalaya-like mountain chains running through what is now the eastern United States.

Rodinia lasted for about 250 million years until it broke up. The process was as messy as any in rock and roll. Different sections of the supercontinent began pulling apart from one another, creating rifts. Some tore all the way apart. The one stretching from the Gulf of Mexico to Illinois and roughly following the current course of the Mississippi did not. Geologists call it the Reelfoot Rift. Google it and you'll also find the names of at least two different bands—one thrash metal, the other bluegrass—but they, too, have gone the way of supercontinents.

Geologists aren't sure why this particular rift began to tear but never fully ripped. What they do know is that pressures from continental forces have mostly closed the rift now; however, it still remains a very fragile piece of real estate. Even though it's in the middle of North America, this area is highly susceptible to what is happening at tectonic boundaries. As strain builds and builds at those margins, faults in and around the Reelfoot Rift serve as a kind of pressure valve, releasing that energy in the form of violent seismic activity. Why here and not the Midcontinent Rift? One theory is that Reelfoot, like Yellowstone and Hawaii, traveled over a hot spot, which melted away some of the crust and left the area particularly unstable.

That instability is made manifest as a network of faults known as New Madrid. We don't know much about them, since the massive river and its sediment-laden floodplain cover them by as much as 200 feet. But we do know that New Madrid is the most active seismic zone east of the Rockies with, on average, about 200 earthquakes a year. Any one of those has the potential to rival some of the biggest quakes this country has ever seen.

What does a quake like this look and feel like? Consider the experience of John Bradbury, a Scottish botanist who plied the very same waters on which the *Island Queen* now cruises. The United States was still in its infancy when the Liverpool Botanic Garden sent him here to gather plants for their collection. Chief among Bradbury's goals was to collect cotton samples with an eye toward improving production strains. He figured the area around New Orleans was as good a place as any to start. However, after arriving in the United States, Bradbury met with Thomas Jefferson, who had just finished his tenure as president. Jefferson recommended Bradbury focus his attention on the alluvial lands around St. Louis. And so Bradbury and his party secured a flat-bottomed boat on which they could live and conduct their studies.

On December 16, 1811, Bradbury and his party passed the town of New Madrid, a tiny outpost located in what is now Missouri. On account of the fault zone, the river makes a hairpin turn there, which in turn made navigation difficult. Bradbury and his party tried to resupply in the town, but found it—and its stores—"straggling" and the source of much disappointment. So they continued onward, eventually arriving at the same bluffs under which the *Island Queen* and I now mosey. There, they spied a small band of Chickasaw with stores of venison. Bradbury and one of his men paddled over in the party's canoe to replenish their meat supply. But the deal went sour when the Chickasaw accused the remaining expedition members, still on their flat-bottomed boat, of trying to shoot their dog (those same party members would later swear to Bradbury that they were merely trying to kill a loon). By the time the situation was sorted out to everyone's satisfaction, dusk was falling. Bradbury and his team made the decision to continue just a bit farther down the river and tie off for the night above Chenal du Diable, or Devil's Channel—a particularly pernicious stretch of rapids and snags created by downed trees. They ate their supper and went to sleep. Here's Bradbury's account of what happened next, which he published in his travelogue *Travels in the Interior of America*:

> I was awakened by a most tremendous noise, accompanied by an agitation of the boat so violent, that it appeared in danger of upsetting. Before I could quit the bed, or rather the skin, upon which I lay, the four men who slept in the other cabin rushed in and cried out in the greatest terror, "*O mon Dieu! Monsieur Bradbury, qu'est ce qu'il y a?*" I passed them with some difficulty, and ran to the door of the cabin, where I could distinctly see the river as if agitated by a storm; and although the noise was inconceivably loud and terrific, I could distinctly hear the crash of falling trees, and the screaming of the wild fowl on the river, but found that the boat was still safe at her moorings. I was followed by the men and the *patron*, still in accents of terror, enquiring what it was.

Bradbury and his team made their way to the stern of the boat, just in time to watch massive landslides rain down from the bluffs. The waves created by those slides almost sank the boat. Hours later, they made their way to shore, only to find that great chasms had been formed by the quake. In a way, they were lucky: At least three other similarly moored boats in the area were sunk by the waves, and all the passengers were lost.

These tremors were forceful enough to be felt over a huge swath of the continent and strong enough, according to the USGS, to "alarm the general population" in an area of 1.5 million square miles—over twice the size affected by the 1964 Alaska quake (which, you may remember, remains our nation's most powerful one—by far). People as far away as New York and Washington, DC, reported being jolted awake by the seismic waves.

Aftershocks continued into the next year. The last one was arguably the most powerful of the three "principal" quakes. It was certainly the most destructive. It obliterated the town of New Madrid and destroyed much of St. Louis. It created immense geysers of sand that covered fields, rendering them unfarmable. Eliza Bryan resided in New Madrid at that time. Her account is at least as colorful as Bradbury's published one:

> At first the Mississippi seemed to recede from its banks, and its waters gathering up like a mountain, leaving for the moment many boats, which were here on their way to New Orleans, on bare sand, in which time the poor sailors made their escape from them. It then rising fifteen to twenty feet perpendicularly, and expanding, as it were, at the same moment, the banks were overflowed with the retrograde current, rapid as a torrent—the boats which before had been left on the sand were now torn from their moorings, and suddenly driven up a little creek, at the mouth of which they laid, to the distance in some instances, of nearly a quarter of a mile. The river falling immediately, as rapid as it had risen, receded in its banks again with such violence, that it took with it whole groves of young cotton-wood trees, which ledged its borders. They were broken off with such regularity, in some instances, that persons who had not witnessed the fact, would be difficultly persuaded, that is has not been the work of art. A great many fish were left on the banks, being unable to keep pace with

the water. The river was literally covered with the wrecks of boats, and 'tis said that one was wrecked in which there was a lady and six children, all of whom were lost.

Apocryphal accounts of these earthquakes have been a part of our nation's collective mythology since they occurred. Ask folks, and they'll probably tell you that these quakes made the river run backwards (it didn't) or that they cracked the Liberty Bell (that's not true, either). We didn't get really serious about thinking about the implication of these quakes until the 1970s, when Otto Nuttli, an unassuming professor of geophysics at Saint Louis University, began collecting and cataloging eyewitness accounts like the ones above. He spent years sifting through microfilm of long-gone newspapers as far afield as St. Louis, New Orleans, Toronto, New York, and Quebec. (If this pastime seems overly laborious to you, spare a thought for the poor geologist Stephen Taber, who assessed the 1920 Los Angeles earthquake by counting broken bottles in pharmacies around that city.)

What Nuttli was looking for was evidence that could place the New Madrid quakes on the Modified Mercalli Intensity Scale: stories of stopped clocks and broken crockery in Natchez, Mississippi; tiny tsunamis in Pittsburgh puddles; portraits swaying in Baltimore. In the end, Nuttli concluded that the quake Bradbury experienced was around an M 7.5, making it about ten times stronger than the one that destroyed San Francisco in 1906. The two subsequent major events included an M 7.3 quake on January 23 and an estimated M 7.5 on February 7. More recent work in paleogeology has borne out this assessment. That means, even today, that these three quakes—which occurred nowhere near a plate boundary—rank as the eighteenth-, nineteenth-, and twentieth-strongest quakes in our nation's history. Take away quakes in Alaska and Hawaii, and they easily make the top ten.

James, our intrepid guide aboard the *Island Queen*, spins a great narrative about these seismic events. He falls for the old "Mississippi River flowed backwards" myth, which can be forgiven since it really seemed that way to eyewitnesses (what actually happened was more like a tsunami wave, though that wave was powerful enough to back up some of the Mississippi's tributaries). But James loses me when he proudly announces to our class that these quakes were easily an M 9.0. At the end of our cruise, when the ladies next to me have finished their wine and packed up their plastic glasses, James greets us at the bottom of the gangway, his hand outstretched.

"Tip?" he asks.

"Sure," I tell him. "Here's one." And then I proceed to explain subduction zones and magnitudes and why 7.5 is plenty strong, especially in the middle of the United States. He smiles indulgently and tells me we'll have to agree to disagree.

Fair enough.

The Midwest doesn't need anything as big as an M 9.0 quake to bring this country to its knees. Otto Nuttli knew that in 1971. It took national emergency managers a couple of years to get on board, but by 1983, we nevertheless had one of the most robust earthquake planning offices in the country here: the Central United States Earthquake Consortium, or CUSEC. The first of its kind, CUSEC was a direct partnership between federal agencies like FEMA and seven at-risk states (Arkansas, Illinois, Indiana, Kentucky, Mississippi, Missouri, and Tennessee). This constituency remained as such for several decades, until an earthquake in northern Alabama suggested it would be a good idea to add that state, too.

Meanwhile, our understanding of the potential for disaster here has only increased. In 1999, FEMA declared four scenarios "catastrophic incidents" for the United States. Such a disaster, they say:

> Results in extraordinary levels of mass casualties, damage, or disruption severely affecting the population, infrastructure, environment, economy, national morale, and/or government functions. A catastrophic incident could result in sustained nationwide impacts over a prolonged period of time; almost immediately exceeds resources normally available to State, local, tribal, and private-sector authorities in the impacted area; and significantly interrupts governmental operations and emergency services to such an extent that national security could be threatened.

So what kind of Armageddon-like event could wreak such havoc? Well, they predicted, probably a major hurricane hitting the Gulf Coast. Surely one clobbering Miami as well. Add to that a major earthquake event in Los Angeles. And finally, you guessed it: one on the New Madrid fault zone, too.

Why?

In part because, again, New Madrid is the most seismically active place east of the Rockies. And, like much of the eastern United States, its cooler, denser rock means that earthquake waves there travel much farther and impact a far greater area (about twenty times greater than an equivalent event in Los Angeles or San Francisco). But the real answer comes back to this notion of hazard and risk. Hardly anyone was around to experience those 1811 and 1812 quakes; in fact, the United States had owned that area for only a few years. And so, while they were obviously tragic for the people who experienced them (and, in some cases, lost their lives or their livelihoods), the nation pretty much kept marching along with business as usual. A lot has changed in 200 years.

In 2009, the Mid-America Earthquake Center, which is based at the University of Illinois, conducted a seismic impact study for the New Madrid region. Investigators and analysts at the center worked with FEMA, the Central United States Earthquake Consortium, and the Army Corps of Engineers to devise a plausible scenario and measure its effects. They settled upon an M 7.7 quake that would rupture along the entire length of the fault zone. That kind of event would result in a lot of the same damage we saw in the hypothetical Salt Lake City quake: 715,000 damaged buildings, 86,000 injuries and fatalities, 7.2 million people displaced, and countless others without electricity, sewage, or potable water.

But unlike Salt Lake City, the New Madrid fault zone really defines itself as a crossroads: a place not just where East meets West, but also where East and West are divided by the nation's most formidable waterways and where nearly all our commerce must pass.

Take DeSoto, the Interstate 40 bridge: a sprawling 3.5-mile arch-and-steel structure. Truckers and barge captains call it Dolly Parton, a colorful homage to its main span, which looks like the letter M if you're a grade school kid and an ample bosom if you're a lonely guy who's been on the road (or the water) for a while.

In 1992, amid growing predictions that New Madrid was ready to rupture again, the Arkansas and Tennessee transportation departments hired TRC, a national engineering firm, to perform a seismic retrofit on the bridge. Making the initial assessment alone was a herculean task: The bridge had been built in 1960, long before any real seismic considerations went into the design and building of our national infrastructure. In fact, prior to the 1970s, most bridges were built to a 1931 code that had only a rudimentary evaluation of seismic strain. Bridges are particularly susceptible to earthquakes: If anything, the asymmetrical design of many older bridges actually amplifies the strain on their supports and decks, at least in part because the different sections of the bridge twist and pull differently, so it's almost as if they're fighting one another. There was also little understanding of liquefaction when their foundations were designed.

We've seen the effects of this oversight in myriad ways. The 1964 Alaska quake damaged every bridge on the Copper River Highway. The Federal Highway Administration estimates that the 1971 San Fernando earthquake did about $100 million in damages to the Golden State Freeway alone. Following that quake, the administration began conducting what it calls "modest" research into national bridge design and seismic hazards. They weren't fully implemented as a national standard, however, until the 1992 Loma Prieta quake, which came with what one California transportation report called alternately "catastrophic" and "spectacular" bridge failures. (That section of the Bay Bridge I mentioned in the introduction to this book was closed to traffic for thirty days after the quake.)

The DeSoto Bridge was one of the very first to be remade in these standards, and its improvements were built with the 2500-year earthquake in mind. The retrofit itself came with a $270 million price tag: more than worth it, the two states insisted, given that over 35 percent of all national trucking commerce passes under its arches each year, and over 30,000 vehicles pass over in a single day. Compare that to New York's Tappan Zee Bridge: It, too, flunked its seismic evaluation in spectacular fashion. In the end, officials there decided that a new $3.9 billion bridge was a better investment.

David DeLeeuw was a resident engineer on the project to rehabilitate the DeSoto Bridge. I say "a" and not "the" because the project took nearly twenty-five years to complete and thus required several generations of engineers working on it. DeLeeuw pointed out the finer points of the retrofit to me from the top floor of the University of Memphis law school building, which itself has undergone a massive seismic overhaul. There, from the top floor of the former customs house, we could watch the dizzying train of cars and trucks speeding over the bridge's six lanes.

"A failure of this bridge would be a catastrophic failure for the country," says DeLeeuw. "Even having it unserviceable for a few days would be an economic disaster."

That includes, he adds, closing it for repairs. The retrofit necessitated adding steel casings to shore up the bridge support, installing expansion joints, and adding pendulum bearings. All of this work, he adds, was completed while the bridge remained open. When really significant work required them to close one lane in either direction of the bridge, they took out ads in trucker magazines announcing the change.

"On the first day of the closure, we showed up to find not a single truck there," says DeLeeuw. "Every last one of them took a detour."

Today, thanks to the retrofit, the bridge can safely move twenty-two inches in the direction of traffic and sway eighteen inches from side to side. That, along with its strengthened foundation, should be enough to keep it fully operable in the event of a quake. Emergency vehicles should be able to use it immediately, DeLeeuw predicts, and regular traffic should resume as soon as officials can complete a safety inspection, which he estimates could take two or three days.

If so, the DeSoto Bridge may well be the only real lifeline in the area. That Mid-America Earthquake Center report estimates that over 3500 bridges in the region will be damaged in an M 7.7 quake, with approximately 1250 rendered impassable. Additional railroad bridge damage will make it even harder to get supplies and services to survivors there, since so many of these communities depend upon neighboring states. To understand why this really matters, consider this: FEMA estimates that the most "at risk" survivors of a New Madrid event will need 15.3 million MREs (meals ready to eat), along with 23 million liters of drinking water and about 62 million pounds of ice, in the first seventy-two hours alone. That's 3500 truckloads. An estimated 1500 search and rescue teams will be needed to look for additional survivors in the region's collapsed buildings. We need a way to get them and their vehicles into the region, too. The DeSoto Bridge could well be closed for all three of those days. Smaller regional bridges will probably be too dangerous to cross, and inspections of them will take even longer. That same FEMA report predicted that "it is unlikely that response workers, evacuees and supplies can be moved from across the river for several hundred miles south of St. Louis, Missouri, to south of Memphis, Tennessee." And so local communities may well find themselves unable to care for those people who survived the earthquake itself.

They won't be the only ones.

Today, the overwhelming majority of this nation's agriculture is grown in the greater Mississippi River basin. Over 60 percent of all our grain exports travel down this river, just part of the over 300 million tons of goods that travel down the Mississippi. Along with grain, we ship coal and petroleum, and everything from copper ore to bottled spirits to gluten to ammonia. Combined, these goods make the ports of New Orleans and south Louisiana two of our country's biggest in terms of total tonnage shipped.

Meanwhile, traffic on this river keeps getting busier and busier. Back in 1811, the first Mississippi steamboat, the *New Orleans*, was actually en route to its namesake city when the 1811 quake struck. It eventually arrived at its destination and ultimately opened up river commerce. With each decade, traffic on this river increased exponentially. So, too, did built communities. The river wanted none of that. Still, we persisted and did whatever we could to make the Mississippi comply.

Like all rivers, the Mississippi is always looking for the path of least resistance. It likes to move, shifting its banks to softer soils, building up sediment and sand bars that eventually become its own natural dams. Its mouth has traveled over 200 miles in the past ten millennia or so. It floods annually, sending its waters rushing across flat deltas.

All of these factors pose problems where human settlements are concerned. The residents of New Orleans realized that straightaway, when nearly annual flooding events wreaked havoc on their city. They began building levees to contain the river in the early eighteenth century. Eventually, those levees pushed farther and farther northward until they became a massive system over 3500 miles in length. Meanwhile, we saw both the wisdom and the challenges of using the river as a major thoroughfare for commerce. But that required a stable, navigable corridor. To get rid of features like the Devil's Channel that had so plagued Bradbury and his contemporaries, we built locks and dams. We also began dredging the river to maintain a channel deep enough for heavy barge traffic. The sum total of all this work created a river that is more navigable, but also swifter and more powerful.

More than one contemporary civil engineer has insisted these attempts were little more than hubris—a modern-day Icarus story. No matter how hard we try, Mark Twain wrote back in 1878, we "cannot tame that lawless stream, cannot curb it or confine it." In attempting to domesticate the river, we are engaging in the prizefight to end all prizefights, and one in which the river will necessarily be declared victorious.

We've already had glimpses into how this might look. The spring of 1927, for instance, brought with it deluge after deluge. Levees from Illinois southward began to fail, inundating communities. In Arkansas, the flooding covered over 6000 square miles. And it rose so fast, people had no time to prepare. About a hundred residents there died in a single day. Untold numbers of mules and cows and pigs did, too. The disaster hit Greenville, Mississippi, hardest of all: Floodwaters killed 1000 people there and left another 10,000 homeless and stranded, prompting a human rights nightmare. With nowhere to go, stagnant floodwaters persisted for months, preventing recovery efforts and encouraging malaria and typhoid outbreaks. By summer, over a quarter million people were still homeless.

If we as a nation took home a message from that tragedy, it was merely that we needed to build bigger levees. And so we did. With the passage of the 1928 Flood Control Act, we reinvested in earthen levees to control our rivers, along with locks and dams to make them navigable. Today, the Army Corps of Engineers alone controls 707 dams, 220 navigational locks, and over 15,000 miles of levees. Each year, they save the country about $150 billion by preventing floods.

The Army Corps of Engineers also maintains 3500 miles of levees as part of the Mississippi River and Tributaries project—a length, they like to boast, that is longer than the Great Wall of China. The levees begin near Cairo, Illinois, and constrain the river all the way down to its mouth. They protect nearly 2 million homes, about 33,000 of which are working farms. They are also the only line of defense against floods that would cripple roadways and vital transportation corridors. At least half their length lies squarely within the New Madrid impact zone.

Routine inspections by the Army Corps of Engineers have established that, in the Memphis region alone, over 464 miles of levees are in "unacceptable condition." That includes the St. Francis Levee, which protects a large portion of the seismic zone, including Memphis. Nearly 400,000 people live behind the protection of that levee. The most recent inspection there turned up a host of deficiencies, including erosion and sloughing, depressions and rutting, animal burrows, seepage, cracking, and issues with slope stability.

John Rice is professor of civil and environmental engineering at Utah State University. He's spent most of his career studying how levees perform in seismic disasters. He says that a rating of "unacceptable" doesn't necessarily mean that a levee will fail in an earthquake, but by their very nature they are some of our most vulnerable infrastructure—particularly where seismicity is concerned. Levees are basically made by piling up dirt and till. They're long structures that traverse a variety of different rock and soil types. Some are built directly over faults. Riverbanks, meanwhile, tend to be composed of soft soils prone to liquefaction. The levees themselves are similarly prone to the phenomenon. In the event of a quake, they can settle to a level that will no longer contain floodwaters. Even a partial failure can result in disaster if it occurs during flood stage.

"A breach is a breach," says Rice. "It doesn't matter if it's ten feet or a thousand. Water finds a way." Besides, he says, the force of water has a way of turning a little breach into a bigger one.

And unlike a lot of dams, these locks and levees were built without redundancies in place. As the seismic hazards in this country change, so, too, does the potential that these structures will fail. Repetitive seismic events, even small ones, can interfere with their structural integrity. Factor in flood-level water, and you have the recipe for catastrophic disaster.

What's the likelihood of an earthquake that occurs during a flood?

"We don't know for certain," says Rice. "But we can quantify the probability that there's going to be a flood and the probability that there's going to be an earthquake. That ought to be enough to make some real improvements."

Rice says he doesn't worry so much about an earthquake that causes a ten-foot breach during, say, August: There's a pretty good chance it could get patched up before the next flood stage.

"But," he asks, "what about that ten-thousand-foot breach? That's going to take a lot longer and a lot more resources to fix. In the event of an earthquake, we may not even have the roads in place to get to it for months. That's what I worry about, come flood season."

That Mid-America Earthquake Center study found that, in the event of a significant earthquake, about 400 of the dams and levees in the New Madrid region would be damaged enough to cause flooding in surrounding areas. Damage would be particularly debilitating in Tennessee, which has the greatest infrastructure in the flood zone. A 2014 Army Corps of Engineers report found that, were this section of levee to fail, a "major disaster" would occur, with property damages estimated to be about $11 billion.

According to that same report, "the most impacted facilities include communication facilities, fire stations, waste water facilities, highway bridges, and schools."

It doesn't take much. During a recent heavy rain, one of Memphis's underground emergency management locations flooded, forcing them to retreat to a mobile command center hosted by the county. That kind of fleet-footed damage control requires officials to have the means to get elsewhere, which is really hard to do if roads are flooded as well.

This problem is not restricted to the birthplace of rock and roll. Across the United States, over 50 percent of our levees have exceeded their life span. Some are over a hundred years old. Levees, especially, aren't built for extreme longevity. Combined, the nation possesses about 100,000 miles of them. They're located in every state and the District of Columbia. FEMA estimates that nearly half the nation's population lives in a county with at least one levee. And yet, says Rice, we have no National Levee Safety Program. Many of our levees have never been formally evaluated. Nearly 85 percent of them are privately owned and under no formal oversight. A lot of them are crumbling of their own accord. The National Committee on Levee Safety estimates that it would cost about $100 billion to rehabilitate our aging system.

"After Katrina, our eyes were wide open," says Rice. "We saw the damage caused by failed levees, and funding became available. But since then, interest has waned. And now funding is harder and harder to get. I worry it'll take another major disaster before we're woken up again."

THANK GOD FOR FEDEX

Inside the passenger terminal of the Memphis airport, a traveler could get the idea there isn't much going on here. The low-slung building opened in 1963, and the interior still harks back to the design aesthetics of that era: narrow halls bedecked in brick and drop ceilings, muted waiting areas with rows of plastic and pleather chairs. In 2013, Delta, which for years had been the anchor airline here, announced that they were pulling out their operations (people in the industry call this dehubbing). The airline took with it 230 jobs and over ninety flights a day. For the first time in almost thirty years, Memphis International was an airport without a hub and, well, the hubbub it creates. In the years since, the airport has diversified its airlines and managed to win back much of the passenger base it lost with Delta, but it's still a much slower place.

I arrived just before 8:00 P.M. on a sultry evening. By the time I entered the terminal, all the shops and restaurants were already closed for the night: Stacked chairs hung from the neat wood tables outside Interstate Bar-B-Que. Heavy gates hung down over the CNBC and PGA stores. In fact, the only signs of life in the entire terminal were two Cinnabon employees finishing their sodas in front of a darkened storefront.

Outside the terminal, on the other hand, it was a mob scene. Literally. A hundred or so angry, panicked, distressed earthquake survivors were clamoring to get into the gate protecting the Tennessee Air National Guard's 164th Airlift Wing Air Terminal Space. Some of them had walked for miles—through the dark, picking their way past ruined roads and toppled buildings. They were desperately thirsty. Others were bloodied and sporting broken limbs. One man was screaming in Spanish—his baby hadn't eaten in over a day. Guard members on the other side of the gate were scrambling to find an interpreter. Others were politely—firmly—trying to quell the riot: As much as they would have liked to, they couldn't help these people at the gate—not yet. Meanwhile, still others were searching for survivors in a downed hangar.

And I? I looked on with utter delight.

It's not that I love chaos and misery. But I do really love first responders—especially when they're practicing scenarios. This one was the first multiorganizational earthquake drill in Tennessee history. At the center of it all was Colonel Raymond Robinson IV. Robinson graduated from the Air Force Academy in 1991, then took a position flying C-17s at Edwards Air Force Base in Southern California. He was there when the Northridge quake rocked the region in 1994. That taught him a lot about how earthquakes work in real time. Over the course of the week I visited, he was working with local, state, and national agencies to anticipate how one would unfold here. He says they learned a lot—particularly when it comes to communication.

Say those Mid-America Earthquake Center predictions are true, he says. Say the bridges and the roads and the levees in this region really leave islands of people unable to get medical help or supplies. At least in those first few days, their well-being is really going to come down to what Robinson's eight C-17s can get in and out of the region.

The C-17 is a nimble workhorse. Boeing, its creator, dubs it "The

Globemaster." It has a huge payload and can land just about anywhere—including on rough runways as short as 3000 feet, assuming the weather is decent (compare that to most commercial jets, which require more than twice that, even in ideal conditions). The first question Robinson has is whether he's going to have access to those 3000 feet—and just how many of his planes will be able to use them.

It's a reasonable question. In the scenario he and his staff are practicing, an M 6.5 quake has rocked Memphis in the wee hours of the morning. Two of Robinson's planes were in their hangars, which have at least partially collapsed. So he's down to just six. It's a Saturday, so most of his staff is at home. Things would be looking much better if it were noon on, say, a Wednesday.

While Robinson is trying to account for his existing crews, Terry Moon, vice president of operations and maintenance for the airport, is working with engineers there to assess the airfield. He says he's pretty confident most of it will be operable in a day or so after the quake. He's less sure about the pipelines that funnel in jet fuel from as far away as Texas. There's a pretty good chance the passenger terminal has sustained major damage as well. Moon is also worried about the integrity of his air traffic control tower. But that may be okay. Some of his air traffic controllers are kicking it old-school—like really old-school. They're out at the base of the runway in an SUV with binoculars and handheld radios. If they can make contact with other regional airports, they can complete a relay that will get those C-17s flying again. That's important: Right now, those planes are the only way to get supplies in and injured people out. They stay fully fueled as a matter of course, so crews can fly them out 300 or 400 miles—beyond the impact of the quake. Once they do, they will become, in effect, an air bridge—a link between helicopters trying to get to those isolated pockets of people and the hospitals and supply chains outside the impact zone.

The trick here, Robinson keeps reminding me, is communication: They've got to be in communication with the airport authority, who in turn has to be talking to the FAA and other towers. They need to be able to talk to local first responders, who in turn need to be able to tell first the state and then the federal government what they need. They need to be able to find those supplies or assets or whatever by asking other states. Probably, phone lines are down. So are cell phone towers. This is the first time all these stakeholders have tried to talk to one another without them.

How is it going?

When I ask Robinson, he's quiet for a few seconds, and then laughs in a low, slow chuckle.

"We're learning a lot," is his eventual answer.

And what's the nature of the lesson?

"Everyone—including us—has to be prepared to take care of themselves first."

That's a hard fact to swallow when there are throngs of people just on the other side of a fence, begging for help.

In this particular case, that angry mob outside the gate is actually composed of some of the best actors in the Tennessee Air National Guard. Truth be told, they had some fun with it. So did the casualties reposing in various postures of agony and distress. In an actual earthquake, says Robinson, the mob is going to be bigger. They're going to be frantic—lured by a lit-up guard base that may be the only beacon in an otherwise dark area. And when it comes to disaster triage, they're not going to rank very high. Robinson says he can't stress it enough: You have to be able to take care of yourself in moments like that—probably for three or four days. Until they get those runways up and running. Until the planes can bring real help.

If you think these runways matter only to the people in the region awaiting relief and supplies, you're mistaken. A 2012 study by the University of Memphis estimated the economic impact of the airport at around $23 billion annually. At least $22 billion of that comes from cargo. Just about every Amazon order you place, every prescription you fill, every part your mechanic needs to fix your car, touches down on this same airstrip. Here, UPS maintains their third-largest hub, which is also their only hub capable of processing both ground and air freight. (Their crown jewel world hub—a sprawling 5.2-million-square-foot complex capable of handling 115 packages a second—is located in Louisville, which is also in the New Madrid damage zone.) But the real elephant in the room that is Memphis International is FedEx. They've based their operations there since 1973 and, in many ways, define the airport.

On any given day, the airport lands about ninety-four commercial flights. FedEx brings in well over 200. Only Hong Kong International Airport is bigger in terms of cargo, and FedEx dominates that landscape, too.

"Thank God for them," says Terry Moon. "Without FedEx, we'd be Shreveport."

The FedEx campus itself would put any Big Ten school to shame: over 880 acres of tarmac, buildings, and a seemingly endless stream of packages.

I meet Rick Armstrong, senior public relations specialist for FedEx, at the FedEx Experience Center just after 10:00 P.M. He's a lifelong Memphis resident who began working as an equipment operator for the company in his early twenties. Five years later, he was one of FedEx's first VJs for HubTV, the company's internal television network. He's perfect for the job: Armstrong speaks with a slow Tennessee drawl and wears a perennial smile. He can talk midwestern baseball but prefers football (especially the Denver Broncos). Each night and into the wee hours of the morning, he oversees the company's incredibly popular Hub Tour.

After shaking hands with Armstrong, I stand off to the side and watch as he offers tonight's scheduled guests coffee and soda and little packaged snack crackers before leading them into a state-of-the-art minitheater, where they watch a quick video on the hub, then head out for their tour. Most of them are account managers for some of FedEx's key customers—representatives from health care and pharmaceuticals, the automotive industry and big-box retailers. Tonight, they include one guy from Africa and another who just arrived from Germany. The latter can't remember how long he's been awake at this point and is really, really glad to accept Rick's offer of coffee.

After they depart, Rick and I join Rick's colleague Jack in a brightly painted company van. Jack is in his late twenties and fairly new to the PR department. He tells me he's just passed his defensive-driving test—a detail that seems peculiar at the time but makes infinite sense a few minutes later as we pull onto the jet ramp. I've never been to Mumbai, but from what I've been told, it's safe to assume that the driving conditions there are a close analogy to what we are experiencing. Those little souped-up vehicles you see pulling luggage trailers at commercial airports? They're actually called tugs, and on any given night, the FedEx world hub has about 1500 of them in operation, zipping along painted pathways and each making a quick two-beep courtesy honk at every intersection. They share the space with giant self-propelling hydraulic lifts and about 150 aircraft, including dozens of Boeing's behemoth, the 777. It's controlled mayhem.

When we climbed into the van, Rick had offered me the shotgun seat so I'd have a better view of everything. Now, as we drive around the facility, he leans forward from the backseat, pointing out the two original hangars, the refrigerated storage area, and more, interrupting himself fairly regularly to offer advice or driving instructions to Jack that go something like "*DUDE!*" or "*GUN IT!!*," at which point, Jack—who is doing an outstanding job of making sure that we don't get run over by a brontosaurus of a jet—will hit the brakes or the accelerator or circle back. I ask him if this kind of driving makes him nervous. Jack shrugs like it's no big deal.

"Just an occupational hazard," he says.

FedEx owns and operates over 650 aircraft, making it the world's fifth-largest fleet (behind the likes of American Airlines, Delta, and United, but well ahead of Air China, Air Canada, and Ireland's Ryanair). Tonight, somewhere between 150 and 180 of them will land on two of the airport's runways. They arrive first from Europe and the East Coast (on account of the time zones there), and then are quickly followed by planes from as far away as the Middle East and the South Pacific. While Jack waits in the idling van and scans the taxiway for more jets, Rick and I climb aboard one 777 just landed from Sydney, where a team of cleaners are tidying up the crew's sleeping quarters and galleys (pilots on these long-haul trips work, sleep, and eat in shifts). Rick stands in the doorway while I take a look inside, where rows of containers, each specially designed to fit the 777's fuselage, about sixty or so in total, are being lowered off the plane. I ask Rick if he wants to step in farther, away from the gusty wind blowing that night.

"No way," he says sheepishly. "I hate everything about flying."

Back in the van, we follow this plane's containers into one of two massive metal buildings known as Matrix East and Matrix West. There, they are scanned, sorted, and redistributed at dizzying speed (about 190,000 an hour) by a system beyond complex. The two matrices include about eighty miles of conveyor belt, and each package is scanned about twelve times as it passes through. Bright yellow feet push boxes off the main conveyor onto one of a series of fun-house slides that bend and corkscrew onto lower conveyor belts going elsewhere. Eventually, each package arrives at another container about to board another jet headed to another destination.

A similar process unfolds in the document-sorting building, which processes 325,000 envelopes and small packages an hour. Rick and I stand there for several dizzying minutes on a metal grated catwalk, watching as each envelope passes by on its own tilt tray, which would eventually send the package down a designated slide and into a waiting bag. Below us, dozens of employees watch each bag as it is filled, then sealed with a printed label and sent on its way. Prior to my visit, I'd been instructed not to wear a skirt on the tour: Standing atop that open catwalk, I finally understand why.

All told, the time it takes to process an envelope or package in the Memphis hub is three hours—less when employees are really on their game (which they usually are). About 1.5 million of them will make their way through the hub every night. If you've ever had blood work sent out to a lab, it may have passed through here (medical containers wear bright orange tape to alert workers). If your doctor prescribed you a medication based on those lab results, chances are it passed through here, too. On a single night, so too will pass thousands of contracts and paychecks, legal decisions, passports, birth certificates, and more. Amazon employs them regularly, as does every major branch of the military. Their biggest client, though, is actually the US Post Office, which regularly asks FedEx to do their heavy (and light) lifting.

Around Christmastime, that number of nightly packages can almost double. Valentine's Day is another peak time: Over three-quarters of the 4 billion commercial flowers Americans buy each year are actually grown in Colombia and Ecuador. But that holiday is eclipsed entirely by the week prior to Mother's Day, which rivals Christmas in terms of peak shipping traffic. Not the run-up to Father's Day? I ask Rick. He shakes his head.

"Nah, not so much. But we do see a little spike in the couple of days after," he says. "I figure it's all those guys returning their ties and underwear."

Overseeing all this scene—and every other hub, plane, and package—is Dave Lusk, senior manager of global operations for FedEx. At least, that's what his business card says. On the grounds of a FedEx facility, he's known as the "master of disaster." He's the guy at the center of contingency planning and triage whenever something goes wrong anywhere FedEx services, which is to say pretty much anywhere in the world. About fifteen minutes before our appointed meeting time, Lusk sends an email apologizing and saying that he will be a half hour late—a big storm is brewing in the Northeast. I didn't think to check my email. And so, a few minutes after that message, he also sent a text. The master of disaster is also the master of multitasking.

When we do meet, I find an outgoing guy in his sixties, with slicked-back white hair and rimless glasses. He's affable and speaks with a twinge of both his Cincinnati heritage and his years in the navy (which is to say his "ma'am" in enunciated with equal parts drawl and military crispness). Lusk worked for three years as an air traffic controller aboard the USS *John F. Kennedy*, a 60,000-ton aircraft carrier that was home to eighty jets and their crews. He liked the gig so much he decided to make a career out of it in the private sector. It can take months and months to get your FAA credentials, though, so he spent the interim working as a flight attendant for Pacific Southwest Airlines. It was an illuminating study in human behaviors and needs, he says. And that, more than anything, comes in handy at his current occupation.

These days, Lusk spends most of his time in the FedEx Global Operations Control Center at a campus adjacent to the airport. It's a room just shy of the *Dr. Strangelove* set and overwhelmingly dominated by a massive movie-theater-size screen (which is actually about forty 40-inch high-tech monitors squished together in a grid). There, he and the Global Operations Control team can track every FedEx air asset in real time. They can keep tabs on weather conditions, congestion, and flight hours.

Lusk explains it is in the nature of the business for things to go wrong.

Sometimes, it's a simple mechanical failure or, in the case of tonight, a violent weather system. When Malaysia Airlines Flight 17 was shot down over Ukraine, restricted airspace around the crash site changed flight patterns and added an extra twenty or thirty minutes to travel time—a major headache when you're managing over 650 planes and airports in every time zone.

Disasters pose a particular problem for a company that needs to be in the air and on the ground no matter what. In the days after 9/11, Lusk had to figure out how to turn airfreight into a ground game. When Hurricane Matthew made landfall in South Carolina, closing airports and roads there, he worked through a FEMA clearing house to get reentry letters for his crews, who were carrying resources for the Red Cross, along with basic supplies.

He doesn't relish them, but Lusk says the nice thing about a lot of natural disasters is that they are polite enough to give advance warning. Lusk's crew includes a staff of fifteen meteorologists who work in conjunction with the company's logistical team. When Superstorm Sandy was bearing down on the Mid-Atlantic, they rerouted fifty planes and stepped up operations in the heartland. When Katrina decimated the Gulf Coast, FedEx had to suspend operations in places like New Orleans and Biloxi. In the meantime, Lusk sent down members of the FedEx security team (who are also commissioned police officers) to account for each employee and hand out cash to those who needed it.

Seismic events, he says, pose a particular hazard that's challenging even for the leader in logistical management.

"Earthquakes are immediate and local. We can't track them. We can't predict them."

So what's going to happen to your medication and your contracts and your bullets and your radioactive waste if and when the big New Madrid quake goes off?

Lusk says you're going to have to be patient—at least at first. His first mission is going to be to make sure all of his employees are safe and sound. He's stashed days' supplies of water and MREs in shipping containers around the FedEx campus, and the facility's twenty diesel generators can keep operations running for about twenty-four hours on a tank of gas. But until he knows his employees have been safely evacuated, your package is going to have to wait.

Back in the 1990s, when the New Madrid fault zone was getting a lot of attention, Lusk asked engineering and seismic scholars from around the region to assess the FedEx facility. On their advice, they've retrofitted the buildings here. He feels pretty sure they'll still be standing, even in a major quake. What he really worries about is fire.

"The fire department isn't going to be here in five minutes. They're going to be digging their equipment out of the rubble," predicts Lusk.

To be safe, FedEx has established a "hot" contingency control center thirty miles east of the airport. "Hot," Lusk explains, as in "plug and play—it's a completely redundant facility." One day each week, his team runs the company's operations from there, just to make sure everything is in working order. Twice a year, they move operations over there for two full weeks.

"We want to really stress that facility," he says. "If we have to deploy there in an emergency, we want to know we've done it before. If you've gone through it again and again, then things start to feel pretty much rote."

The company also maintains a duplicate data center in Colorado in case of this kind of disaster—part of what Lusk calls "risk mitigation by mitigating risk over geography."

While that switch is happening, Lusk's other major concern will be to get his aircraft out of Memphis. That'll have to wait until the runways have been inspected and tower controls are fully functional. As soon as they are, Lusk says his plan will be to get his planes to underused hubs in Indiana or South Carolina that can take on the extra sorting pressure.

"This is the philosophy we use in any event—severe winter storm, riot, anything," Lusk explains. "We compartmentalize the event. It's like a tourniquet: tie it off, keep it localized."

Every month, a scheduling team updates all routes and flight times for the company. One of their main tasks is to envision a month where the Memphis hub is down—or destroyed.

"It won't be painless," predicts Lusk, "but it'll work."

I posed Lusk's plan to Jim Wilkinson, executive director of the Central United States Earthquake Consortium. Wilkinson's office building also sits in the shadows of the Memphis airport, though, unlike FedEx, you can walk its entire expanse in thirty seconds or less. Wilkinson cut his teeth with the Mississippi Emergency Management Agency, where he worked on population protection—particularly in the event of an earthquake.

He says he feels confident that companies like FedEx and UPS, along with Amazon and other high-volume distributors, would weather a New Madrid event pretty well.

"I'm not saying they're not going to feel the hurt until they're fully up and running again," says Wilkinson. But, he adds, their contingency plans and ability to shift people and productions around the country—or even the world—will mean the overall impact will be manageable.

"It's the little guys I really worry about," he says.

In July 1997, a huge rock slide shut down Interstate 40 near the Tennessee–North Carolina border. It was supposed to be the height of tourist season. Instead, everything from mineral spas to whitewater rafting companies stood mostly empty as traffic was rerouted through neighboring towns for three months. A similar event occurred in 2009. The Appalachian Regional Commission found that, during both disasters, the hospitality industry there lost anywhere from 50 to 80 percent of its business. Sales in local restaurants and shops declined by as much as 90 percent in some cases. Even the local hospitals lost as much as $200,000 a month in revenue.

That's what really worries Wilkinson. He points to towns like Charleston, Missouri, located at the top of that state's boot heel, just off Interstates 55 and 57. It's a humble town of about 5900 people with all the economic downturns that have blighted the midwestern United States over the past decade. The two major manufacturers in the area—Noranda Aluminum and Westvaco paper—have collectively laid off nearly 1300 people in the past eighteen months. The town has tried to make a go of an annual Dogwood-Azalea Festival, but most of the local revenue still comes from interstate traffic—folks jumping off the highway for a steak and a baked potato at the Glenn restaurant or some vintage teacups at Rocking Chair Antiques. They're the ones that can't make it two or three weeks without any traffic, says Wilkinson.

Karen Teeters is the executive director of the Charleston Chamber of Commerce. I asked her what earthquake closures of the interstates would mean for her town. "I'd say it'd be comparable to the nail in the coffin," she told me. She points to places like Reeves Boomland, a 60,000-square-foot fireworks purveyor. They employ seventy-five people. The Flying J Travel Center has at least thirty-five on their payroll. These are hourly workers. Without traffic, they're out of a job—and a salary.

She says her community has been lobbying for years for seismically sound bridges linking them to Kentucky. Their proposal even includes secure rail traffic so that supplies can be sent that way, assuming highways and the river are out of commission. They haven't had any luck so far. When that Mid-America Earthquake Center report talks about bridges being down for months or even years, they're talking about places like Charleston.

Seismologists at the University of Memphis Center for Earthquake Research and Information say the area is about thirty years overdue for an M 6.3 quake. As for a repeat of the 1811–1812 events?

"It's certainly within the realm of possibility," says Chris Cramer, a professor and researcher there. We were chatting back at the base of the I-40 bridge—the one that may well be the only bridge left standing in the region, were such a quake to occur. Cramer admits the likelihood of that quake occurring in the next decade is low, but says he doesn't have a lot of patience for people who say those events are fluke-ish or not repeatable.

"The science I can tell you about based on recent studies has shown again and again that the New Madrid seismic system is accumulating stress," he tells me there. "It is preparing for the next series of earthquakes. You can't wish that away by simply making an argument that it's not going to occur here. There are structural reasons for earthquakes to occur here. And they're going to again."

Cramer says he can't predict whether these quakes might rattle the area in one year or a thousand. And that, he says, is all the more reason to take action now.

"This is about using foresight. We have the time to prepare. If these quakes are hundreds of years off, we're just going to get that much more on our investments."

Jim Wilkinson agrees.

"Hiding behind the idea of an 1811-type event waiting five hundred years to come around again is a very naive way of looking at hazard," he warns. "Mother Nature has a way of throwing probabilities out the window and surprising us. Over and over again, earthquakes occur outside of everything we think we know about seismological science. It's incumbent upon all of us to be as prepared as possible for whatever the roll of the dice brings."

He'd like to see a shift toward proactive plans for sustainability. One of his very favorite is a "disaster-resistant community," a concept first introduced by the International City/County Management Association. Evansville, Indiana, is our nation's first pilot, or "showcase," community. There, all community members have committed to large and small steps to minimize the effect of a devastating earthquake. Some of the steps are easy—securing computer monitors and bolting down hot-water heaters. Others involve grant-funded disaster recovery business alliances, intended to find strength in numbers and ensure that all business owners end up still on their feet. The town, in turn, has pledged to beef up its flood-mitigation programs, which will decrease homeowner insurance premiums for all its residents.

"It's going to take all of us to stay up and running," says Wilkinson. "You can have the most seismically secure building in the whole wide world. But if the people around you don't, you're going to be out of business anyway."

THE LUCKY FRIDAY MINE

If we learned anything from the SAFOD core, it's that, to really understand earthquakes, you have to get down into the earth. Childhood dreams of digging to China may be impractical, but the intuition behind such plans is good. To understand just how fragile and capricious, just how unstable, our planet really is, it's best to see—and feel—it up close.

I set out from Moscow, Idaho, for the state's Coeur d'Alene region just before 3:00 A.M. And so I missed much of what is so beautiful there while driving in the sleepy dark about 150 miles south of the campus of the University of Idaho. French-speaking fur traders stumbled upon the area centuries ago. After discovering there were no easy deals to be made with the Native American tribes who called this region home, the traders dubbed it Coeur d'Alene, or "hearts as sharp as an awl." Despite the ominous moniker, white prospectors continued to visit in search of their fortunes. Some of them were certain they had found just that in 1878, when gold was discovered in the region. Investors soon realized that mining for a combination of silver, lead, and zinc (the three metals are often found together) was much more profitable. Zinc, for instance, is used to galvanize steel. China was once biggest consumer and producer of the metal, but the shutdown of several of its biggest mines has left the booming country with a zinc deficit. The whole world needs more silver, which is used in every laptop and cell phone out there. Finding new sources is getting harder and harder. At one time, ninety mines dotted the mountains here in Idaho's panhandle. Now, it's just a handful.

One of the largest is Lucky Friday. With a shaft that goes down over 9587 feet, it's also the deepest in the United States—and one of the most studied. I'd reached out to Jeanne DuPont, the company's communications director, and asked if I could visit. She wrote back just a day later to say yes. What transpired over the next week was a series of emails that, taken out of context, could be considered pretty incriminating. What was my shoe size? The circumference of my chest and hips? Questions like these can get a company in a lot of trouble. Unless, of course, that company needs to provide gear for a science writer who wants to spend hours and hours over a mile underground.

My instructions from Jeanne were to arrive at the mine by 5:30 A.M. for my mandatory safety orientation. Aboveground, Lucky Friday is a massive campus of industrial buildings, all made of corrugated metal and painted the same pale green. Overlooking it all is the five-story shaft tower, which looms over the rest of the complex. As instructed, I parked in a gravel lot and walked up a steep flight of stairs covered in the same corrugated metal. Just inside the main building, a dozen or so miners were milling around. Each one wore tan work pants and a forest green T-shirt with "Lucky Friday" emblazoned on the front and back in reflective lettering. This, I quickly learned, makes it much easier to find someone when the only illumination you have is a headlamp. Most of the guys were also already wearing their safety helmets, headlamps, and noise-canceling ear guards. Some also wore packs strapped to their belts that would monitor air quality for the day.

The first of the day's only two elevator trips down the mine shaft was still over a half hour away. It's the only way to get into the mine, and the elevator doesn't run again for eight hours. In other words, there is no such thing as being late for work at a mine. Miss the elevator, you miss a day of work. And once you are down there, you're not coming back up until the night shift. So guys tend to arrive at work early, then hang out and talk about the same sorts of things guys at work talk about anywhere in the country. The morning I was there, that mostly consisted of Little League standings and fishing plans for the weekend.

Like most mines, Lucky Friday runs a day and night shift. And so, twice a day, a horde of miners enter the main supply room and confirm their assignments with a guy behind a counter. Then they suit up in their safety gear, fill up their enormous Igloo thermoses with ice water, and move a metal dog tag bearing their assigned employee number from one board to another. This is a practice as old as the mine and utterly essential in terms of keeping track of who is underground. Miners wear a matching tag on their safety gear. The idea is that it will survive an explosion or cave-in and make it possible to identify that miner in the event of a disaster. The miners I met are affable, fun-loving guys. But to the last, they approach this part of their day gravely.

In my email instructions, Jeanne had instructed me to ask for Nick Furlin, who'd be handling my safety orientation that morning. Walking in, I wondered how we'd find each other. Turns out, I needn't have bothered: Even in the twenty-first century, a woman walking into the assembling morning shift tends to get noticed—especially, it turns out, when she's carrying a notebook and clearly has no idea where she's going. During my twelve hours or so at the mine, I saw two other women, but they were both tucked behind desks in portable trailers. And so it didn't take long for word to get to Nick that I had arrived. He appeared, shook my hand with a warm "Hey, how you doing," and then we were off.

Every year, the Mine Safety and Health Administration hosts a national rescue contest, where teams from around the country compete in written tests and real-life scenarios ranging from first aid to rescue. It's a lot like the Olympics for mining. Nick Furlin has been on the Lucky Friday team almost as long as he's been at the mine. When I arrived at Lucky Friday, the competition team had just placed second overall in first aid and third for team tech, an event where two technicians must work to ensure that self-contained breathing apparatuses are in working condition.

Those devices are crucial to a miner's daily routine. I spent my first forty-five minutes at the mine proving I could work one. And that was when I first began to really rethink this idea. I've always had an issue with claustrophobia. When I was a kid, I'd never sit inside a ferry, no matter how long the ride. I hate driving in tunnels. Since 3:00 A.M., I'd been trying to ignore the fact that I was about to spend the day thousands of feet underground navigating tunnels that, in some cases, were only inches wider than I am. Nick predicted it'd be the heat that got to me before the claustrophobia. He gave me my own giant Igloo thermos full of ice water and warned that some places in the mine get up to 130 degrees. Workers there must spend mandatory time in cooling rooms to keep their core body temperatures in the safe zone.

I did pretty well for most of our orientation. I'd swapped out my cowboy boots for steel-toed ones. I wore a lantern and emergency radio clipped to my safety belt; I felt secure in my company-issue boots; I was even okay with the dog tags on my helmet and neck that would be used to identify me if I was crushed down below. But everything changed when we got to that CO_2 converter. In the event of a cave-in, the device is designed to turn carbon monoxide into carbon dioxide, which takes longer to kill you. It's a complicated chemical process, involving compounds like potassium superoxide and lithium hydroxide, that generates an incredible amount of heat in addition to sequestering breathable oxygen. The devices are also packed with talcum powder or cornstarch to prevent the moisture in a miner's breath from clogging them. Inhaling it, warns the CDC, creates "resultant coughing" and "a gag reflex." Miners, warns the instruction manual, "are advised not to remove the mouthpiece but to cough directly into the unit."

To get down the 7500-foot mine shaft, I'd have to prove I could assemble, don, and deploy the converter in under thirty seconds. I could. That felt good, too. But then Nick made me promise I wouldn't remove the mouthpiece in an emergency, even after the effects of the conversion seared it to my lips. "It'll hurt like hell," he told me. "We're talking third-degree burns. But you'll probably live long enough to get to real oxygen."

That's when things got real.

I didn't have a lot of time to think about it before Nick introduced me to Ben Henderson, the mine geologist who'd be taking me around all day. Barely thirty with a big red beard and an even bigger smile, Ben is enthusiastic about most things. But nothing gets him quite so excited as mines. As an undergraduate at Western Washington University, he played rugby. But it was his time in the geology lab that he really loved. He wrote his thesis on fluids and chemical reactions in rocks and says he'd always imagined he'd be doing this work.

Ben and I joined the line of miners walking across a gravel lot to the shaft itself. Each carried an oversized thermos, along with a nearly identical plastic cooler. We packed into the elevator, six people across, six people deep.

Anyone—or anything—going in or out of the mine must use this one elevator car. To get a tractor or a pickup or even a dump truck down there, you first have to disassemble it into pieces that fit in the elevator. In the case of a new Dodge pickup truck the mine recently acquired, that required a blowtorch and sledgehammer.

Our trip just required smooshing together. Social formalities don't really work in that kind of situation, and so instead the miners made a few quick jokes as we descended. Then they asked prying questions about why I was there in the first place. When I mentioned seismicity, the elevator got quiet. These guys spend ten hours at a stretch in narrow tunnels they've created. Most often, they're working alone, burrowing deeper in mineral deposits, spraying the walls of their tunnel with water to keep down the heat and dust. A miner's partner may be a quarter mile away. In each case, their tunnel is following a mineral vein deposited by water millions of years ago. By their very nature, these veins follow fault lines.

The miners know these faults by name: The bigger ones go by monikers like Lewis and Clark. Osburn. Dobson Pass. There are hundreds of smaller ones, too.

Mining and seismicity almost necessarily go hand in hand. Whenever material is extracted, preexisting stress is redistributed. Sometimes, it increases and accumulates over time. It can get released in anything from a microfracture not even perceptible on a seismograph to a significant quake. Each year, around the world, seismographs in mines record tens of thousands of events (Germany's tiny mine seismograph network, which includes just two remote stations, recorded over 1000 in 2016 alone). The USGS maintains a catalog of what it calls "routine United States mining seismicity," which it defines as "routine explosions and planned roof collapses at mines and quarries in the United States," and includes regularly updated information on fifteen states, including Alabama, Minnesota, and New Mexico—places hardly considered seismic hot spots.

The catalog used to be called "Probable Mining Explosions in the United States." I'm guessing nobody liked the uncertainty that implied. Regardless of what you call it, the catalog includes some noteworthy events, including an M 4.2 event that occurred at an Alabama coal mine in 1986, one of the bigger seismic events experienced by the state in recent history. The strongest quake included on that list is an M 5.3 event at Wyoming's Solvay mine. On the morning of February 3, 1995, an area almost a mile square collapsed, creating an M 5.3 quake that released millions of cubic meters of methane and ammonia. Over fifty miners were belowground at the time. Most of them came out with only minor injuries. Two became trapped after they lost their way in the maze of debris. Rescue workers reached them two days later, but one of them suffered a major heart attack as they struggled to get him back aboveground. He died at the hospital later that day.

No one at the USGS could tell me why events like the ones at Solvay are included on the list of "routine" mining events. The event certainly wasn't planned. And the residents in neighboring communities who had their mobile homes overturned by the blast certainly didn't view it as a normal course of events. For months, the event itself was reported as a natural one: Newspapers and even many geologists reported that it was an earthquake that had prompted the collapse rather than the other way around. A scholarly article published later that year also viewed the event as a natural one. It wasn't until MSHA conducted an extensive investigation with the help of the US Bureau of Mines and regional experts that it was determined that the mine collapse was responsible for the seismic event.

In the Coeur d'Alene region, understanding that seismicity is a particularly complicated enterprise. The entire area is being squeezed by tectonic activity. It creates a lot of horizontal pressure in the region. Lucky Friday became known internationally in the mining industry after it figured out how to deal with that pressure. It's a process they call "underhand cut and fill mining," which disperses force while maintaining the integrity of open areas for miners. Spiraling out of the main shaft of the mine are ramps, which look a lot like large tunnels. Off each ramp are individual slots. Off each slot is a cut—a much smaller tunnel about ten feet high and twenty feet or less wide. That's where each miner spends the bulk of his shift.

There's a comforting regularity to a day's shift in a mine like Lucky Friday. Each day, a miner returns to his cut and mucks out the broken rock blasted by the previous shift. He'll then install a series of bolts to stabilize the cut, drill in a little farther, and add more bolts. He'll spray it all down with water to keep it cool. The rock he breaks will get scooped up by diesel-powered loaders and haulage trucks, then taken to an ore chute, which will churn it aboveground for processing.

Once a cut has been exhausted, workers fill it in with a combination of cement and sandy waste known as mill tailings, then begin again above or below it.

When a new section of the mine opens up, miners make bids on where they'll end up and with whom. There's a lot riding on this decision. Miners with the most seniority get to pick both their cuts and their teams. They receive bonuses based on productivity, so an underperforming cut or lazy team member can mean a difference of thousands and even tens of thousands of dollars in an annual paycheck. A miner might be in a single cut for two or three months, slowly drilling the same narrow tunnel until he reaches the end of a vein. In that amount of time, they see stuff. A lot of it sounds crazy.

Each time a miner deepens a cut or bolts in safety reinforcement, he is redistributing stress not just in the mine itself but also in the surrounding geology. That stress gets stored up in much the same way it does as the result of tectonic movement. Sometimes, it is dissipated as the rock settles into new configurations. Sometimes, just like at plate boundaries, the stress in and around mines is released in violent fashion. Even through their ear protection, miners will hear the kinds of booms created by thunder or the Blue Angels. Sometimes a wad of rock will shoot out of a cut wall like a cannonball coming through the hull of a wooden ship. Or, out of nowhere, a huge cloud of dust will appear and black out their headlamps.

During our elevator ride, one miner tells me he's seen waves on the mine floor that make it look like it was made of water. Another says he's seen the walls of a shaft bow and then get sucked back in as if the place were possessed by a poltergeist.

Ben believes them. When we step out of the elevator at 4900

feet, he shows me a disused railroad bed. It used to be straight. Now it . . . meanders.

Our first stop is a machine shop populated by two mechanics having their morning coffee. Aside from the walls of rock, it is easy to believe you're at any number of similar offices staffed in reasonable places like a city's public works department or even the maintenance office of a school district. Two guys sitting in well-used office chairs. A cluttered metal desk. Fluorescent lights. Even the obligatory pinup calendar. We chat. The longer we do, the more I begin to notice signs that this is no ordinary office, like the yellow vinyl air ducts pumping cool air and the very, very dark tunnel that leads out of the space.

We hop onto Ben's four-wheeler and head out through miles of tunnels. Within seconds, it's apparent to me that I could never get back to the elevator on my own, no matter how much my life depended on it. Ben is my lifeline. Without him, I am lost. We covered this during my safety orientation. This far underground, there is no way to communicate with miners, save for a few emergency landline phones scattered throughout the shafts. The only way to notify miners down below of a problem is by a practice known as "stenching" the mine. In the event of an emergency, workers aboveground begin pumping huge amounts of ethanethiol, a highly noxious gas, into those yellow ducts. That smell when you have a gas leak or turn on a gas stove without lighting it? That's ethanethiol. It is incredibly pungent: Humans can detect its odor at rates of one part per 2.8 billion parts of air. Even at its lowest level, the smell is a trigger for our olfactory senses. A polite way to describe its smell is "rotten eggs" or "cooked cabbage." That's at the one-per-2 billion-part ratio. I've been told that the rate at which it gets pumped into the mine is such that I'm going to start projectile vomiting before I'm even really aware that I'm smelling it. Fun.

The point, though, is that this is the signal to miners that they need to make themselves as safe as they possibly can—as quickly as they can. For some, that may mean getting back to the elevator shaft with the hope that it'll be running. For others, it's finding one of the refuge stations interspersed throughout the mine. These safe havens are reinforced and have enough food, water, and medical supplies for a few days or a week, depending upon how many miners are using them. If Ben has a heart attack or is somehow otherwise incapacitated, I'm supposed to make my way to one of those stations. I have absolutely no idea how I could accomplish this. I try to seem light as I jokingly ask Ben about his health, but I definitely listen hard when he answers.

The miners I meet do the same.

Mines only really make a profit when they are efficient. Most of what is done there is of no value to anyone. It's only the veins of metal that count. Ben's job is to figure out where those veins are and how to get at them. Doing so is a combination of old school and new. We spend most of our day driving from ramp to ramp. It is dark and hot and dusty. But mostly dark. And surprisingly quiet. When I ask him what he loves about life down here, he turns off the engine of the four-wheeler. It's true: There is nothing as black or as quiet aboveground. The absence of light almost has mass. My eyes don't know what to do. In just a few seconds, they start trying to invent light—a flicker here, an orb there. Once they give up, I can see that there is a special kind of peace down here.

As Ben starts the four-wheeler back up, he asks if I get claustrophobic. I tell him it seems a little late in the game to be sussing out that sort of thing. He laughs and asks me if I'm afraid of heights. I soon understand why.

To collect samples from one section of the vein, we must climb a couple of hundred feet up a very narrow shaft. The ladder is made out of two-by-fours and carpenter's nails. Each time I lift a knee, I bang it into one of the rungs. My back slides against the rock behind me—that's how tight a squeeze it is. We slither into an old cut. Ben takes something that looks like an old pickax and hammers out a couple of rock samples, which he places in a canvas bag, labels with a paper chit, and then slings over his shoulder. After a couple of similar stops, he begins to look like one of the seven dwarfs. He's a pretty good sport when I speculate aloud about which one.

Once he's aboveground, Ben will analyze all of these samples and plug the data into a complicated software suite. In general at Lucky Friday, the mineral vein runs first to the northeast and then turns sharply south. Figuring out where exactly all of this happens is Ben's primary job. The samples he takes belowground, along with 3-D imaging and topographical maps of the area, allow him to keep track of the mineral veins belowground, in terms of not just their location but also their structure and condition. He'll provide all this information to the mining engineer. Together, they'll decide where a cut should run and for how long.

Most of our day is spent visiting active cuts to take samples. Every so often, we pull off on a ramp and stop the four-wheeler, then hike up or down to an active cut. Ben and the miners maintain a complex relationship. He's college educated and, by mining standards, has a pretty cushy job. Down below in the mine, he looks different. Whenever Ben shows up, the miners have to stop what they're doing and pull back long enough for him to chip out a sample. That takes away from the day's yield. But letting him do his job also makes it more likely they'll be able to stick with the vein, which is good for their paycheck.

To the last, every miner we visit is almost courtly in his formal reception of me. They stop whatever they're doing, turn off their equipment, take off their heavy gloves, and shake my hand. They stand with me in pools of warm, murky water as Ben takes his samples, and make polite conversation. They love that I want to know about their work. We talk about kids and ex-wives and horses and how long they've been coming down here. Most of them are second- or third-generation miners. They're proud of that. And they know their job comes with a lot of risk.

Five years before my visit, a huge section of rock—about ninety feet long and twenty feet wide—sheared off in the Gold Hunter section of the Lucky Friday mine. Larry Marek, 53, and his brother were working together on the night shift. Larry began the shift by hosing down a pile of muck right below Gold Hunter. It collapsed an hour into his shift. In an instant, he was buried in thirty feet of rock. It took rescuers nine days to recover his body. In a subsequent investigation, a geologist from the MSHA concluded the rock that fell did so in part because it was separated by two faults. When the pillar above it was removed, it slipped down. The judge who heard the civil case surrounding the accident likened it to trying to hold up an unabridged dictionary by pressing it in your hand between two paperback novels: The bigger tome is bound to slip out. He ordered the mine to pay $280,000 in fines.

That year, 2011, was a particularly bad one for Lucky Friday. In addition to the collapse that killed Marek, another miner was crushed to death while trying to dislodge a jam. A fire prompted a full evacuation of the mine. A few months later, a seismic event injured seven. Amid growing safety concerns, the Mine Safety and Health Administration shut down the mine until repairs could be made. Meanwhile, Hecla, the corporation that owns the mine, hired new management.

The mine itself reopened in 2013. Since then, officials have been very transparent about their safety procedures. And the mine is one of the most studied in the world. Thanks to that work, geologists have a much better understanding of just how culpable *we* are for initiating earthquakes. Studies at Lucky Friday, which have been replicated in mines around the world, reveal that the overwhelming majority of tremors there don't actually occur because of the faults themselves. Instead, they happen because our activities redistribute stress in the rock. Weak sections of rock may have rested peacefully for millions of years, since there was no stress placed upon them. But as we carve out sections of the earth or just plain alter it in any way, we send a kind of jolt of lightning through the rock that didn't exist there before. The weak rock, now taxed, shifts, slides, crumbles, or explodes. The more we create and move that stress, the more likely it is that weak rock will fail.

Ralf Fritschen, a German geophysicist, studied this phenomenon in a coal mine located in the southwestern corner of that country. Not only did he find that the extraction of coal was enough to create strong earthquakes, but he also surmised that even seemingly innocuous activities could set off major events. In one instance, the simple act of cutting a new road in the mine was enough to induce a series of damaging earthquakes. The strongest one, an M 4.0, caused "considerable damage" to buildings on the surface. That one was enough cause to shut down the mine entirely, though the quakes continued to persist days afterwards.

It's a powerful lesson in just how fragile—and volatile—the ground beneath our feet really is. Most of us forget this if we ever even knew it. Miners can't. Their lives depend upon a constant awareness that what we think is rock solid is unpredictable and alive.

That's what Bob explains to me while he and I wait for Ben to finish taking a sample of his cut. Bob's been working this mine for thirty-five years. He tells me, without blinking, that his son-in-law, a former collegiate all-American football player, died in a neighboring mine just a few years back. He was just shy of thirty and working with both his dad and his cousin when it happened. What killed him—and what injured those seven miners back in 2011—is known in the mining industry as a rock burst.

The MSHA defines a rock burst as a "violent failure of rock" that results in at least one of four eventualities: It forces the evacuation of a miner, it blocks a passage, it impairs ventilation, or it disrupts mining for more than an hour.

Miners around the world have reported rock bursts for well over a century. The first recorded induced seismic event in what is now the Czech Republic was catalogued in 1872. Similar activity has been reported in France's Provence region, the gold fields of India, and Japan's undersea coal mines—some of which have been active for over a century and all of which are places I would recommend to the next James Bond villain looking for a secret lair. Germany began installing seismographs in its coal mines as early as 1908, the same year South Africa began its first formal study of mine-induced seismicity. The latter country remains the leader on the subject, in part because most of the world's deepest mines are located there, including Mponeng gold mine, the deepest, at nearly 13,000 feet.

The deeper mines got, the more seismic events like rock bursts became a serious problem. In the United States, they were virtually unheard of until the twentieth century was well under way. One of the first reported ones here occurred at Arizona's Miami copper mine on April 17, 1913. The day before, miners had complained about hearing cracking noises in the mine; they warned the management it sounded like an entire section of the mine might calve. Supervisors decided to keep men out of the area where the noise was loudest, but that wasn't enough. The next day, a huge explosion of rock shot out of one of the mine walls. And, just like at Hebgen Lake, this rock event was precipitated by an air vacuum, followed by hurricane-force winds, as 3.5 million cubic feet of air was sucked from one side of the mine to the other. The force of that wind was strong enough to throw men and equipment around the mine. When the shift boss heard the rock burst, he commandeered four workers to accompany him on a rescue mission. Rock continued to rain down there, and all four of the workers abandoned the boss. He went back up aboveground and grabbed four different men. They, too, refused to go any farther when they saw the conditions below. Eventually, he found two miners willing to help him get some of the injured miners to the surface. Ten of them were rushed to the area's newly opened hospital. Five died there of their injuries—mostly skull fractures.

Less than a year later, residents of Houghton, Michigan, confused another rock burst—this time, at the Quincy mine—for a naturally occurring earthquake. Tremors from the blast were recorded at Cleveland's St. Ignatius College, part of the first Jesuit seismic network. Beginning in the 1930s, the US Bureau of Mines began developing a network of listening equipment specially designed for mines. They began testing it at Coeur d'Alene's Sunshine mine in 1941, after two miners were killed in a rock burst in June of that year. A more codified version of the network wasn't used for another twenty years, and even then, those networks were focused on monitoring individual sections of rock rather than the mine as a whole.

In 2007, Utah's Crandall Canyon mine, located on the Wasatch Plateau, also witnessed a catastrophic collapse—in this case trapping six miners underground. Rescuers toiled for four weeks to free the miners. Three of those rescuers died themselves during a second cave-in. Eventually, the decision was made to call off rescue efforts and seal the mine forever. A year later, seismologists at the University of Utah released a scholarly paper arguing that, once again, it was the mine that had created the quake. Their data came largely from Bob Smith's seismic network.

Further compounding the issue was the federal decision to disband the US Bureau of Mines in 1996. Oversight was transferred to the National Institute for Occupational Safety and Health, a division of the CDC, but little of the equipment or expertise made the switch. In fact, by some estimates, over two-thirds of the USBM's programs were canceled and about 1000 employees lost their jobs.

Still, as mines get deeper, seismic risk gets more substantial—even where there aren't any major known faults: Again, our underground activity is substantial enough to mimic tectonic forces. A magnitude 5.6 earthquake was caused by a German potash mine in 1989. And scientists say that, in places like the Wasatch region of Utah, that kind of mining activity may prompt even bigger quakes. Scientists at the University of Utah's seismograph network say that upwards of 97 percent of the seismic activity in that region is created by mining-related activities: In fact, they recorded over 19,000 events in a thirty-year period there. Lucky Friday once experienced a rock burst of M 2.5 or greater about every three months. That was before they opened their new, deeper shaft. There's still no word on what the rate will be once operations begin there.

Meanwhile, other events like bursts in coal mines continue to be a problem for miners around the world. Those in the coal business refer to these seismic events as "bumps." Scholars documented 172 of them in the United States over a sixty-year period and say they led to eighty-seven fatalities and 163 injuries. Most of these bumps occur in central Appalachia—states like West Virginia, Virginia, and Kentucky. Colorado is another hot spot for them, as is Utah. In 2000, a team of scientists from the National Institute of Occupational Safety and Health installed a network of seismic monitors in one such Utah coal mine. Over a five-month period, they recorded 13,000 seismic events, including an M 4.2 quake that sent rock slides plummeting onto highways and blocking train tracks. Scholars also believe that a coal mine in New South Wales, Australia, may have prompted an M 5.6 earthquake just three days after Christmas in 1989. Unlike its neighbor New Zealand, Australia is considered a nation at low seismic risk. This quake was centered near Newcastle, an industrial city that distinguishes itself as the largest coal-exporting harbor in the world. The quake rattled much of the downtown and collapsed the Newcastle Workers Club, killing nine people. Four others died in similar building failures, and over a hundred people were injured. More than 60,000 buildings were damaged in a swath that extended as far as Sydney, over a hundred miles away. The cause? Excavation in a large coal mine there. Those same scholars also speculate that mining practices could have similar effects on other unknown faults in the area.

That's what happened in Wilkes-Barre, Pennsylvania, in February of 1954. Subsidence caused by a coal mine there caused what looked and felt like an earthquake. Hundreds of houses there were

damaged, and the shaking sent residents running into the streets—
or at least trying to. That same shaking tented sidewalks in bizarre
angles and buckled roads. It also broke gas and water lines and sent
a plume of methane into the air.

I spoke with Amy Louviere at MSHA about the safety of US
miners underground. She said she only had statistics for coal mines
and that those statistics only included reported events. She also said
the number of bursts in the nation's coal mines has been on a steady
decline over the last thirty-five years, pointing out that there were
twenty-five incidents reported in 1983 but fewer than five in 2015.
That's true, but what has happened in the intervening years is a
little more complicated: lots of statistical peaks and valleys.

What is definitely true is that mining companies are becoming
more adept at dealing with these incidents. Today, on average, the
number of mines that must close because of induced seismicity is
very low—about one every five years. And since 1983, fifteen miners have died as a direct result of mining-induced seismicity. It's a
number that's been on the decline in the past fifty years, but also
one we'll probably never see get down to zero: So long as we are
altering the earth, it is going to move in unexpected and sometimes
violent ways.

Rock bursts aren't just limited to mines proper. They also plagued
construction of ultradeep intake tunnels at China's Jinping II Dam.
Unlike at more conventional dams, the intake tunnels at Jinping run
nearly 8000 feet below the surface of the earth. The deeper construction workers dug, the more common the rock bursts became—over
500 observable ones in the dam's four tunnels. Geologists there had
some luck predicting when they would occur by employing microseismic monitoring systems. They began drilling "pilot tunnels" to
try to relieve stress before the intake tunnels themselves were begun.
Meanwhile, in Peru, design engineers have spent a hundred years
trying to figure out how to divert water from rivers on the eastern
side of the Andes to the country's drought-ridden Pacific coast. The

eventual solution was the Olmos Trans-Andean Tunnel, the second deepest in the world. Drilling began in 2008 but was soon halted when crews experienced over 16,000 rock bursts. Normal containment methods were no match for the seismic events. Eventually, they had to install elaborate roof support systems to contain the blasts.

Switzerland's Gotthard Base Tunnel, which is both the longest and deepest railway tunnel in the world, experienced similar events during its construction. They were further compounded when construction crews there came across a massive, previously undiscovered fault directly in the tunnel's path. Over a hundred earthquakes were observed during just two years of the tunnel's construction. Over forty-seven of them occurred in just a single twelve-hour period. And most of those, ironically, were centered on what was to be the tunnel's emergency stop station. The whole scene prompted one particularly adamant Christian blogger to dub the project "Lucifer's Swiss Tunnel." That wasn't enough to sway the Swiss. They opened the tunnel for operation in early December 2016. After the inaugural train ride through the thirty-five-mile tunnel, the chief of the Swiss National Rail Service emerged to a flock of reporters.

"It's Christmas," he told them.

Not quite.

DIGGING BACK EAST

It was a glorious late October Sunday, and New York City was a mob scene. At 145th Street and Broadway, police had long since given up on keeping the crowd under control. Families—over a thousand of them—were camped out on every available vacant lot, eating picnic lunches, killing time, causing trouble. Peanut vendors, sandwich men, and opportunistic kids were selling everything they could get their hands on. Boys taunted cops. Cops taunted boys. It was, wrote one newspaper reporter, "a typical Coney Island sort of a crowd." Except, of course, that it was much bigger. And it kept growing.

Three days prior—a Thursday—the city's board of aldermen had convened in their chamber. The mayor called upon a resident bishop to offer a benediction, and then asked the city's chief engineer to make a brief statement. Then, the entire delegation left city hall and disappeared down a flight of street-side stairs: highly irregular behavior in 1904. For the few lucky reporters allowed to join the delegation, what happened next was nothing short of revolutionary. With the turn of a silver key, Mayor George McClellan launched the city's first subway system. The plan was that he would then serve as honorary engineer for a short hop to the next stop. As it turned out, he had so much fun he stayed on until the end of the line, then rode a second train back again.

New York's subway wasn't the world's first—London had that distinction. It wasn't even America's first—that was Boston. But it was arguably the grandest, and it certainly came with the most fanfare. By the time the weekend rolled around, an estimated 350,000 people were waiting in line with the hopes of hopping a ride. For most passengers, it was an utterly affecting experience. An eighteen-year-old man was so overcome with the experience of being underground that he fainted straightaway and had to be taken to an uptown hospital. He opted to take the elevated train back home. Another panicked passenger leapt onto the train with such force he fell and broke his arm.

Most riders avoided the need for medical attention but were no less flummoxed. This was the first time any of them had ever been underground, and it was an utterly arresting experience. For years, many of them had seen the construction of the subway, much of which was undertaken by digging enormous ditches through the city that then would contain the tubes and required infrastructure. One observer described it this way: "And here is this great new hole-in-the-ground, stuffed with one knows not how many potential reactions on the life and the look of the town. And yet every day we ride over miles and miles of it with scarcely more than a languid musing as to the likelihood of dynamite explosions, or a peevish interest in magic devices by which contractors manage safely to support the pavement over which we ride, the L structure, or whole sheaves of underground pipes." Horses regularly fell into these man-made caverns. Burglars were said to use them to escape the police. They stopped traffic and rerouted pedestrians and generally made the city a hassle to navigate. Meanwhile, millions and millions of cubic feet of dirt were taken from the cuts and trundled down to the bay in awkward carts.

From a technical standpoint, none of this was really all that revolutionary. And there were some pretty significant snafus, of course. An explosion of blasting powder at 41st Street killed eight people and injured dozens of others. In the tunnel just below 169th Street, miners hit a fractured section of rock that calved into the tunnel, killing five of the workers there. At just about 37th Street, a cave-in took out not just the tunnel under construction but also several brownstones nearby. The tunnel in question was about sixty feet underground and through a particularly dense section of rock. Miners were in the process of blasting it when a huge section faulted and slid "as a card slips out of a loosely shuffled pack." It took with it first the stoops and areaways of the homes. I'll let our dandy observer tell the rest of the story:

> Parts of the front walls soon followed, and the crowd of idlers and nurse-maids and delivery-boys who gathered a few minutes after the first cave-in enjoyed the delectable experience of gazing into the very heart of each house, just as you look at an interior on the stage. One gentleman was in his bath-tub at the time. His valet burst into the room. "Quick! quick! You must get out of here, Sir!" cried that worthy. "There's been an earthquake, Sir, and the house is falling in!" "Indeed!" observed the gentleman with interest, and he finished his bath. He dressed himself, and loading his film camera and lighting a brier-wood pipe, he sallied forth, and when his wife's mother arrived on the scene from a distant part of town, whence she had driven at breakneck speed to save her child, she found her son-in-law standing on the brink of the chasm in front of his door-step, pointing down into it a film camera, the shutter of which

he was working with the liveliest enthusiasm and de-light. This teaches us that a bucolic equanimity may be preserved even on a metropolitan street beneath which a tunnel is building, and that nerves may be suppressed even in New York and in a somewhat neurotic age.

For all the supposed neuroses of the age, it was also one marked by real geological ingenuity. Consider what was happening under the East and Hudson Rivers. There, miners known as sandhogs worked to drill through a miserable combination of very hard rock and very soft soil. Rock bursts were common, and conditions were often unmanageable.

Take the Steinway Tunnel, one of the first to connect Long Island and New York City. It proved both expensive and deadly, bank-rupting the first two companies that attempted to complete it. In 1891, William Steinway, of piano company fame, agreed to fund the project, figuring it would boost the value of his real estate in-vestments on the island. But flooding and constant rock bursts soon slowed construction to a snail's pace. Meanwhile, attempts to blast through bedrock shattered shop windows and cracked the founda-tions of apartment buildings. To try to break through the rock, workers used ample amounts of dynamite. On a particularly cold morning in December 1892, they arrived to find that the cartridges had frozen together overnight. The foreman on duty ordered one of his workers to thaw them out using a steam box. There, the whole lot of it detonated, killing five and injuring dozens of others—including local residents. The resulting court case bankrupted the tunneling company and again stymied construction. Water con-tinued to flood the tunnel. By February 1893, Steinway and his company had given up on the prospect of besting New York's sub-terranean geology and boarded up the unfinished tunnel. It would be another fifteen years before the project was completed under new management.

Like many tunnels built in this era, the Steinway Tunnel was originally intended to shuttle trolley cars full of commuters between Queens and Manhattan. As such, it is impossibly narrow and near archaic with its cast-iron tubing. Today, it serves as a key thoroughfare for the 7 train, which leaves from 1st Avenue and 42nd Street, then quickly dips down some fifteen stories and into the tunnel itself. It's a short ride to Citi Field, home of the Mets, from there.

The tunnel itself is a perennial pain for transit employees. It's so narrow that they can't even work in the tunnel when trains are running—there's no room for them to get out of the way of an oncoming car. It presents even more complications for emergency managers—especially those concerned about seismic hazard.

When the Steinway Tunnel finally opened, over twenty years had passed since the 1884 earthquake. Geologists still insisted it was a fluke—that it'd never be repeated. Surely it couldn't be topped by a stronger one. And so no seismic-loading calculations whatsoever were made during the tunnel's construction. In fact, no seismic considerations whatsoever went into its design and construction.

That worries people like Vince Tirolo. Tirolo has spent most of his professional life studying the seismic hazard of these tunnels. When he was an undergraduate, it never occurred to him that such a profession even existed. But then he fell madly in love with a young woman studying at a nearby Catholic college. They wed as soon as she graduated, but he was still a few credits shy of a degree. To make ends meet, he took a job at the transit authority while finishing up his degree in night school. Through chance and happenstance, the city assigned him to the division specializing in underground transportation. He's barely come up for air since. These days, he divides his time between academia and work as a consultant both for the city and for the various transportation companies that manage underground tunnels. And, he says, if you really want to understand the integrity of these systems, you first have to understand how they were made.

In the heart of Manhattan—or deep into New Jersey, for that matter—tunnel construction looked a whole lot like what happens in any of our nation's mines: Workers there could pretty much just blast through the bedrock and end up with a reliable pathway. But the closer they got to the rivers themselves, the less effective and the more dangerous this process became. There, the ground shifts abruptly from solid rock to a mucky, unreliable combination of silt and sand. To penetrate it, workers would use what's known as a tunnel shield: a kind of rocket-shaped capsule made of steel about twenty feet in diameter that would protect the miners as they inched their way through the sediment, first carving it out and then laying down the tunnel lining. Under the rivers proper, they'd pump in highly compressed air to stop water from flooding the shield and drowning the workers. Because of continued concerns over decompression sickness, workers moved quickly and didn't necessarily stop to check their work. However, Tirolo says, a much bigger danger arose when those shields hit the boundary of sediment and bedrock. The latter made it next to impossible to advance the shield, so miners would gouge out the rock by any means possible—and as quickly as they could. Along the way, they left craters and gaps—voids where the tunnel lining isn't in contact with the earth around it. Tirolo says these gaps are as big as three to four feet in places. Workers tried to fill in the gaps with grout or gravel, but even if they succeeded in filling the holes, they probably didn't do so with the kind of material that could stand up to a major quake: Tirolo worries a quake today would cause those tunnels to rattle around. Many of them were cracked or dinged during construction (there's no graceful way to transport enormous cast-iron tunnels, and it's even harder to get them belowground without bumps and nicks). That damage, coupled with the rattling effect of the tunnel lining during any seismic activity, could well result in a major issue—especially where the bedrock meets softer soils. During an earthquake, those soils stand a good chance of liquefying (there's that word again) or, at the very least, becoming compacted by the shaking. The effect of either would be to drop that section of the tunnel while the portion in the bedrock remains in place. Hold one end of a Popsicle stick while you pull down on the other end. Eventually, that stick breaks and you end up with a lap full of melting cherry ice. Cast iron is heartier than your stick, but it, too, has a breaking point.

That's what happened in 1995, when the Hyogoken-Nanbu earthquake ruptured near Kobe, Japan. It crumpled the columns of the Daikai subway station, causing an entire section of the ceiling to collapse. A subsequent investigation found that the station, which was built in 1962, had also been designed without any seismic considerations.

This is not necessarily a justification to forswear the subway in favor of Uber. In fact, being in a tunnel underground is one of the safest places to be in an earthquake. In 1976, a major one rocked Tangshan, China, killing nearly 250,000 people. When the quake hit, thousands of miners were belowground in the region's eight coal mines. Almost to the last, they survived unharmed. In 1999, central Taiwan witnessed a magnitude 7.3 event known as the Chi-Chi earthquake. Aboveground, over 50,000 buildings were completely destroyed; even more were severely damaged, including nearly 1000 schools. Over a hundred major bridges also sustained major damage, as did multiple highways. Belowground, business continued as usual: The highway tunnels were unscathed, save for some minor cracking in the tunnel lining. The Taipei subway was similarly unharmed.

Studies have proven again and again that underground structures tend to incur less damage than those aboveground. That's in large part because systems like subway tunnels are encased in rock and soil that will help hold them in place—unlike surface structures that can sway and rattle without any impediment. There are some caveats, of course. Conditions like those surrounding the Steinway Tunnel will always be problematic in an earthquake. Tunnels laid close to the surface are likely to suffer many of the same problems seen aboveground. Tunnels and facilities carved out of soil tend to be less stable than those resting in bedrock. Reinforced tunnels do better than those that aren't. That's also true for tunnels that are lined, provided that lining is well established. Symmetrical tunnels do better than asymmetrical ones.

If your preferred mass transit route is on the right side of those dichotomies, you're probably going to be in luck, even in the event of a major quake. Take San Francisco's pride and joy, the BART. On an average day, it carries about 360,000 passengers, nearly half of whom pass under the San Francisco Bay in one of the longest immersed tube tunnels in the world. During rush hour, as many as 6000 passengers are often in the tunnel at one time.

The BART system was built in the 1970s. Designers knew there'd be earthquakes and that the soil, especially on the San Francisco side, is lousy. The tunnels' ventilation system includes specially designed joints that allow it to rock and roll without any damage. After the Loma Prieta quake, it continued with round-the-clock trains and became a real lifeline across the bay in ways that the damaged bridges couldn't.

Still, as hazard forecasting has become more sophisticated, the risk of an earthquake to the San Francisco area has increased. The USGS now says the likelihood that the Bay Area will see a major seismic event in the next fifteen years is around 62 percent. In scientific terms, that means you should put your money on the quake. Probably, they say, that quake will be more intense than Loma Prieta. And there's a good chance that it will happen much closer to San Francisco. So while the subway system may have performed just fine before, there's a decent chance it won't go unscathed in this hypothetical quake. Key among engineers' concerns is the possibility that the soil under the bay tunnel will liquefy, making the tunnel itself unstable. Meanwhile, some of the joints of the tunnel are already reaching their capacity for movement, which means they'll be unable to weather additional strain. New seismic analysis has revealed that an additional tunnel actually crosses directly over the Hayward Fault, widely considered one of the most likely places where the next earthquake will originate. Given the kind of displacement we've seen in past San Andreas earthquakes, it wouldn't be unreasonable to assume one there could rend the ground on either side of the tunnel by several feet or more, taking sections of the tunnel with it.

In 2000, BART launched a massive billion-dollar earthquake safety program. Once completed, it will have retrofitted thirty-four stations and seventy-four miles of track, along with tunnels and utility systems. City officials are hoping to have the work completed by 2022. They're also hoping the quake won't have happened by then.

It's a different story back in New York. Vince Tirolo recently published a paper outlining his concerns regarding those iron tunnels and their performance in a quake. I asked him what kind of a response it received. "Not much," he told me. "That's kind of a funny thing."

I asked the same question of the Metropolitan Transportation Authority, which oversees the subway system. I told them that I'd talked to Tirolo and representatives from the city's Office of Emergency Management and wanted to talk to the person who handles seismic concerns for the MTA. They told me they couldn't accommodate my request. I asked why. I'm still waiting for a response.

—◠◠◠—

At the Horsebrook Cafe in Plainfield, Connecticut, I order a plate of chicken stir-fry and position myself at the bar under a stuffed coyote howling at an invisible moon. Most of the other décor is an homage to Beantown athletics: a couple of neon Patriots signs, an obligatory Red Sox poster, and a Celtics schedule. I arrived in town after football season ended and before baseball began, so the lone TV is tuned to the Food Network. Guys, mostly wearing leather jackets and vests (several of which are emblazoned with the American flag), who were well habituated to the Sunday game look at the screen awkwardly as two sets of women battle to make themed cupcakes for the launch of a new line of sneakers aimed at teenage girls. It is a far cry from second and goal.

Plainfield is a gritty town. But it's also a close-knit one, where people keep an eye on one another. The police blotter for the week I am there includes a couple of burglaries and one report of a domestic assault. That same week, cops also make the news after they arrest a guy for pulling kids in a sled behind his truck at high speeds: The reader comments after the story mostly attack the cops for not having anything better to do.

At the Horsebrook Cafe, an older man wearing a Yankees hat (an audacious move in these parts) stops by when my lunch plate arrives. He's never heard of stir-fry before and, after eyeing my vegetables, says he'll stick to Budweiser. Sitting next to me is a young guy who introduces himself as Dane. During the winter, Dane installs grain silos throughout New England, but that's mostly just to make ends meet. His real career is managing Plainfield's Dairy Bar. He has a nine-month-old daughter whom he loves a lot. They, along with his girlfriend, live next to the Dairy Bar on one of the town's busier streets. After he asks what I'm doing there, he tells me I'd be much better served in writing a book about why cars keep crashing into his house. He has a point.

Prior to a few years ago, Plainfield hadn't had a single earthquake— at least, not one since it was settled in in 1699. Then, in October of 2014, three tiny ones rattled the area. No one thought much of them until the start of the New Year, when these little quakes became a daily occurrence: over 180 of them in about six months. Dane says he definitely felt them, and that his buddy's house has cracks in the drywall caused by all the tremors.

Seismologists refer to this phenomenon as an earthquake swarm. Sometimes, they turn out to be lots of foreshocks heralding a much larger event. Other times, they just fizzle out. We still can't predict with any accuracy which way a swarm will head.

John Ebel, senior research scientist at Boston College's Weston Observatory, was asked to consult on the Plainfield quakes. He told residents there that he had no way of knowing whether they might be precursors to a major event.

"Something strong could happen, but it's probably a low probability," he told the local news anchor on a widely watched segment about the quakes.

Residents didn't find a lot of comfort in that kind of prognostication. They'd already heard and felt a lot of seismic activity that spring. They described the quakes as thunder and explosions. Each time one of the quakes ruptured (and, curiously, they were almost all in the morning), dozens of Plainfield residents would call 9-1-1. As the quakes continued, residents became increasingly anxious. So the chief of police and board of selectmen decided to take the only action they knew how: They called Alan Kafka, associate director of Weston Observatory, to speak to the town. He arrived, as promised, on a freezing Friday night. He'd expected a couple of dozen people, maybe a few more. Instead, he found the auditorium packed with over 1000 people. Some had already seen their homes damaged by the quakes. Others talked about how their well water turned murky brown with each tremor. Still others stayed at home, saying they worried the big one might happen while they were all packed into the auditorium.

Kafka is an earnest, patient guy. He likes talking science with the general public. He wears well-worn jeans and equally well-worn button-down shirts, usually with a bright T-shirt underneath and a couple of ballpoint pens in the pocket. He has a gray beard and wire-rimmed glasses and looks a little bit like a scruffy Mikhail Baryshnikov. He is eminently approachable. That night, he brought with him an information-packed presentation backed by a Power-Point slideshow. He planned to spend the first half of it explaining the science behind seismicity in the eastern United States. He didn't get very far.

What he wanted to say was this: Like Idaho, the Eastern Seaboard is in the middle of a tectonic plate. Or at least it is now. About 400 million years ago, the western edge of what is now Africa slammed into the continent, part of the process that formed Pangaea. That collision definitely left a mark. Two hundred million years later, the two continents were torn apart. That created plenty of strain—and faults—as well. One of the most noteworthy in Connecticut is the Lake Char–Honey Hill fault zone. That's an abbreviation. The actual zone comprises two faults: Honey Hill and Lake Chargoggagoggmanchauggagoggchaubunagungamaugg. Be glad you're not a grade school kid there. Lakes like that make spelling "Mississippi" a cakewalk.

Connecticut has dozens of other faults, but seismologists don't know much about them, since they've always believed they are dormant. There's one exception to that belief, and that's the town of Moodus, about thirty miles west of Plainfield. "Moodus," like "Char," is an abbreviation. The original name was Machimoodus, so designated by the three Native American tribes that originally occupied the region. "Machimoodus" translates loosely as "land of bad noises." During the seventeenth century, they assumed the rumbling growls that seemed to emanate from a cave there were manufactured by a tetchy god. They happily sold the land to Puritan settlers, who purchased it for thirty coats. That seemed like a great deal until they, too, heard the groaning. They figured Satan was responsible.

You can still hear the noise if you time your visit to what is now a state park just so. I did my best on the way to Plainfield. It was a gray and pouring day. All I heard was raindrops on my umbrella.

Back at the town meeting there, Kafka told the crowd he didn't know if the Moodus noises were related to what was happening in Plainfield, but they were definitely proof that the area has active seismicity. He began to explain more about this when he was interrupted by a woman in the audience. She already knew about Moodus and ancient plates and all of that. She wanted to know why her town was suddenly plagued with daily earthquakes.

"Do you know why this is happening in Plainfield?" she asked. "What's going on?"

Kafka had to admit that he didn't know. It was a powerful moment for him. He and I talked about that over a beer. He's long become accustomed to the uncertainty inherent in his field. It's the pursuit of information that really excites him. In the first days of the swarm, he and some colleagues from Weston installed an array of monitoring stations around Plainfield. They were able to locate the epicenter of many of the quakes: directly below an abandoned quarry. He presented some of his early findings at a regional seismology conference in Memphis. Several of the audience members asked him about the correlation between the quarry and the earthquakes themselves. It was no longer mined, but it certainly had been for some time. And there were several other active quarries in the area. Could they be responsible?

Precedents certainly exist. In 1974, an M 3.3 earthquake was centered on a limestone quarry in Poughkeepsie, New York. Twenty years later, an M 4.6 earthquake shook southeastern Pennsylvania. It was the region's largest ever recorded and opened a giant sinkhole measuring 200 feet square. It also caused over $2 million in damage and prompted 9-1-1 calls in New York City and Baltimore, where residents felt the shaking. The epicenter was located at a limestone quarry. After mining there ended in late 1992, the quarry was flooded. Little quakes began there almost immediately. A neighboring quarry remained in operation. Regional seismologists speculated that either activity could have been enough to prompt the 1994 quake: Maybe, for instance, the removal of the limestone was enough to shift geological strain, just like miners have seen belowground. Maybe flooding the quarry increased pressure enough to prompt a seismic event.

Something similar occurred in Taiwan, during the construction of the Taipei 101, the tallest building in the world. Prior to the project, seismicity had been negligible in the area. But as contractors added story after story to the building, that began to change, first with microseismic events and eventually with two earthquakes felt by many in the area. And, as we've already seen, something as simple as just blasting can create its own seismicity. Just a few years ago, a quarry outside Chicago set off an explosion that registered 3.7 on regional seismographs. It was strong enough to rattle houses in the region and knock things off shelves.

At that Memphis conference, several seismologists in the audience stood up to say they thought the location of the quarry at Plainfield was too much of a coincidence to be overlooked.

Margaret Thomas, the state geologist for Connecticut, agrees.

"The correspondence is quite striking," she says. "The surface locations of these events swirl in and around that quarry."

That makes perfect sense to her.

"When you have a quarry operation, you do a lot of blasting. So you blast, it cleaves off the walls of the quarry, you pull those off and crush them or whatever. But blasting over time also causes the rock to fracture—that's why you do it. But it also imparts other stresses into the earth. If it weren't doing that, you wouldn't have the rock fracturing."

The quarry in question produced crushed stone, primarily for concrete and asphalt. It began blasting at least as early as 1982 and was in operation until about 2013. That's at least thirty years of fracturing—maybe more. Thomas concedes that even that kind of sustained activity is rarely enough to spark an earthquake swarm, at least in the Northeast. Instead, she says, it was probably an unusually large amount of water that then seeped into those cracks that really got things rolling. She and her team went out to the quarry just after the first swarm began. It was full of snow.

"Once you have the fractured rock, you increase infiltration of surface water," Thomas explains. "Normally that's a good thing. We want increased infiltration of surface water. That's where our drinking water comes from. And a lot of runoff can do all sorts of other damage."

But, she says, too much seepage might have been enough to lubricate already stressed faults underground. And that particular winter, there was a lot of melting and refreezing, which would have caused water to seep in at high rates and then freeze, expanding and putting additional pressure on the already damaged rock (Ebel thinks this might have been why most of the earthquakes happened right around dawn—that the cold nights prompted the expansion).

This idea that groundwater could be responsible for earthquakes is over a century old. As early as 1912, geologists began to speculate about various causes of earthquakes, particularly those related to the hydrological cycle. In 1913, Robert W. Sayles, curator of the Harvard Geological Museum, published his treatise "Earthquakes and Rainfall" in the *Bulletin of the Seismological Society of America*. In it, he compared precipitation levels and seismic activity around the world and concluded that a clear correlation existed: When the former increased, so, too, did the latter. Sayles's article received precious little attention from the geological world, and he didn't help his case much in 1925, when he contended that a series of earthquakes in New England the previous winter had been caused by a lack of rainfall. Still, says John K. Costain, professor emeritus at Virginia Tech, Sayles was on to something.* In the late 1980s, Costain pioneered the idea of hydroseismicity, which contends that water cycles—including tides—are responsible for a lot of intraplate activity. He points to the idea that much of the earth is cracked and

---

* Since my interview, Professor Costain passed away.

fractured down to depths of ten miles or more. Depending upon the permeability of rock types, water can fill and move through those cracks, changing pressures. Even the slightest increase, he says, is enough to trigger an earthquake. Over the course of three days in August 2005, for instance, the Swiss Alps saw record rainfall. About seventy-two hours after that weather event, they also witnessed a swarm of some forty-seven earthquakes.

Costain focused most of his work in the Central Virginia Seismic Zone. There, he noticed an uptick in quakes after hurricanes or torrential rains: After Hurricane Camille brought a deluge to the region in 1969, for instance, a series of moderate earthquakes followed, including one measuring M 4.6. After Hurricane Irene passed the region in 2011, seismologists noted a significant increase in aftershocks associated with the Virginia earthquake of that same year.

Costain also found that increased stream levels could spur an increase in seismic activity, and while he can't prove it indisputably, he wonders if flood events on the Mississippi could be responsible for similar spikes in activity in the New Madrid Seismic Zone as well.

Costain, Ebel, Thomas, and Kafka all agree we'll probably never know for certain what caused the quakes in Plainfield. Seismologists in the East—and particularly in New England—haven't been able to find any clear connection between the few faults mapped there and the earthquakes that occur. The limited number of instruments in the region means that earthquakes are hard to track, and most of the faults in the area still lie undiscovered.

Kafka says that a lot of people would rather hear that a big new fault has been discovered in their backyard than hear the kind of uncertainty he has to offer most of the region. But, he says, at its heart the geology of New England is ambiguous. Earthquakes happen there all the time. Maybe it's because the ground is just now beginning to bounce back from the glaciers. Maybe it's all that old strain from our collision with Africa. Maybe it's us. Could be all three.

The citizens of Plainfield remain frustrated that more definite answers are not available. Kafka remembers one polite but outspoken woman in particular who, during his presentation, reminded him that she was not one of his students—she was a resident trying to figure out how to stay safe.

"People don't necessarily want the complex science," says Kafka. "They want to know if they're going to die."

He says that realization was a real wake-up call for him. "I'm looking at these interesting infrastructure records and the audience is wondering if and when they're going to survive."

That really got him thinking about how earthquake science is communicated. He concluded his talk at the conference with some real challenges for the other seismologists in the audience: "To what extent do seismologists say and do things that are genuinely helpful to the people concerned about the earthquakes? How much does the basic research that scientists do address the public's concerns?"

No one had an answer.

OUR INGENIOUS WELLS

The town of Stillwater, Oklahoma, has two grain elevators, four pawn shops, seven BBQ joints, and approximately 900 earthquakes a year. The businesses have been around for decades; the earthquakes have not. Tremors were first felt there in 2006, the same year oil companies like Devon and American Energy began aggressive hydraulic fracturing campaigns in the area. Mark and Ilke Crismon have lived just outside of Stillwater since their home was built in 1971. At first glance, little has changed in the intervening years. There's still the same wood paneling and white painted brick, a metal roof, lots of sliding glass doors. Outside, their land sprawls for about 160 acres, past a man-made pond and some discarded farm equipment, along with one of the largest vegetable gardens you've ever seen. Inside, animal pelts and a framed picture of a Native American riding a painted pony bedeck the walls, along with a reproduction of the *Mona Lisa*. Mark and I sit on their back porch in metal patio chairs—the kind that bobble-rock when you lean back in them. His is burgundy. Mine is turquoise. Separating us is a large stump Mark has fashioned into an end table. There's a turkey feather wedged in it. Scattered around us are empty sunflower seed hulls, detritus from the bird feeders Ilke has hung in every possible spot on the porch.

I'm here because I contacted Angela Spotts, cofounder of Stop Fracking Payne County. I told her I wanted to see what was causing the huge spike of earthquakes in Oklahoma and Texas. She drove me to meet the Crismons.

Mark is seventy-six and the living embodiment of both Johnson-era politics and country folkisms. He wears his full white hair slicked back in the style of an early Bob Barker. His button-down shirt is neatly creased and worn above a sparkling white undershirt, which just peeks through the collar. The shirt itself is flannel and tucked into a pair of high-waisted faded jeans. He wears work boots that extend well up his calf but only laces them partway. He chain-smokes, knocking ash of his cigarette and into an old white pitcher as we talk. Ilke wears jeans as well, along with a worn tan fleece. Her hair is also slicked back but much thinner than Mark's. She's seventy-five and battling cancer. She stands, leaning in the sliding-glass doorway, mostly listening. Ilke is a Republican. Mark is a Democrat. They tend not to talk too much about politics.

Below Ilke's feet—below Mark's and my feet—is a web of cracks in the concrete and foundation. Each one was caused by a shallow earthquake in the last five years. In a way, the Crismons have it easy: They can still keep their prints hanging on the walls, their glasses and vases in a highboy. Residents in neighboring communities can't. Some of them just have nails and faded wallpaper where their paintings and family pictures once hung. They've taken to using only plastic cups and plates: Their glass and china ones just broke too easily.

The Crismons have cracks in their living room, too: mostly in their fireplace and on their slab floor. There's a pretty good one in the foundation and down the length of the driveway as well. Mark once tried to get them fixed: He filed a claim with the company that holds his earthquake insurance policy. But by the time they took out his deductible, the check they sent was only for about $1000. Mark was so mad he wouldn't even cash it. He figures he'd have better luck throwing dice at the Cimarron Casino down the road.

Two faults bisect the Crismon property. As best as anyone can tell, they've been there for millions of years. For most of the two centuries that white settlers have occupied this territory, no one even noticed they were there. That all changed around 2008, when Mark and Ilke started hearing tiny pops—like maybe kids were shooting air guns or firecrackers in one of the fields. But there were no kids around. And the pops got bigger. They turned into claps and rumbles. And then they became booms—really, really loud ones.

The Crismons both have more experience than the average American when it comes to weathering explosions. Ilke was five and living in her native Nuremberg, Germany, when the Allies decimated that city with bombs at the end of World War II. She says she still has something that feels a whole lot like PTSD, even after all these years. Mark served in Korea. They both say what's happening at their house rivals any trauma in their past.

I look at the cracks underneath Mark's chair as they say this to me. I take in the bird feeders, the pastoral scene before us. Mark and Ilke experience daily tremors—sometimes as many as seven in about that many hours. The tremors wake them up at night. I ask them to describe what it's like. Mark kicks one of the wood pillars holding up the porch roof. "KA-BOOM!" he bellows. "That's a 3.4."

But it's not really. A few minutes later, I'm sure a bomb has gone off across the road. And then another. I duck. Literally. Like flinch and then roll up into myself, as if making myself small will somehow make me safe as well. What was that? I ask. No one else seems fazed.

"Earthquakes," says Mark. "Small ones."

He, Angela, and I move to his war room. It's actually the corner of a machine shed, set a hundred yards or so off the side of the house. A couple of old refrigerators make a half-wall defining his office space. Corrugated metal walls make up the other two. He's lined that metal with particleboard in some places. That wood is covered in animal tails—russet and ruddy, some with a white tip. I guess coyote. He corrects me: white-tailed deer. He offers me a seat in an old recliner and a cup of peach brandy he made himself. On top of an old metal office desk sits an industrial-size purple Folgers can Mark uses for his ashtray. It's all decidedly old-school, except for the shining new laptop installed there by Oklahoma State University. On it, a series of advancing lines in red, blue, and green report any seismic activity in the area. They look like an EKG of someone who has just had a heart attack.

The OSU laptop is connected to a seismograph in Crismon's yard, one of about a dozen installed in the area by the university in recent years. From it, Crismon has a god's-eye view of the world's inner workings. He knew about the deadly 2015 quake in Nepal before some seismologists did.

But that, of course, is not why the seismometer is there. It's there because the Oklahoma landscape is changing in dramatic and maybe even dangerous ways. What makes that hard to bear for people like Angela and the Crismons is that these changes are not only our fault; they're also entirely avoidable.

These three Okies say they understand their state's fate has been tied to gas and oil since long before it was a state. About 500 years ago, the area that is now Oklahoma was first inundated by a shallow ocean. Those waters brought with them a whole host of prehistoric animals and plant life. When the waters receded, the organic material remained, buried in sediment. The sea returned again and again, bringing with it more material. This process continued until the land began to look very much like an elaborate layer cake.

Those organic layers, once compressed, formed enormous pockets of oil and natural gas, often encased in even larger plumes of seawater. The ground there has been leaching oil and gas for at least as long as any human has trod it. In the nineteenth century, the Comanche saw these areas as places of healing and fed the oil to their children and animals. The Chickasaw burned continuous torches fed by the expulsion of natural gas. The neighboring Cherokee tribe wasn't initially interested in either oil or gas, but they did want a ready supply of salt so they could preserve their food. To find it, they'd drill down hoping to find brine. In 1859, they also found oil.

By the 1920s, Oklahoma wells were producing over half a million barrels of oil a day. The natural gas also produced by these processes was considered a huge inconvenience, creating explosive and uncontainable gushers. When workers found gas, they usually burned it off.

Demand for oil dropped dramatically during the Great Depression and sent the state's economy spiraling (the Dust Bowl didn't help matters much, either). But it rebounded during World War II, and Oklahoma soon found it was unable to keep up with demand. Oilmen dug their wells wider, then deeper. When that didn't work, they tried to crack the rock itself, hoping there was still oil to be leached. First they tried corrosives like sulfuric acid. Then they tried forcing sand into the wells with enough velocity to crush the stone that acid couldn't eat away. On St. Patrick's Day 1949, Erle Halliburton sent a team of his employees to a well in Duncan, Oklahoma. There, they pumped oil and 150 pounds of sand into the well hole—propping agents, they called the slurry. Production increased 75 percent, and the treatment cost Halliburton a mere $900. He decided to patent the process, which he dubbed "hydrafrac." By the end of that year, he had 332 fracking operations under way. Five years later, Halliburton had more than 3000.

Whether he knew it or not, Halliburton was banking on the very same phenomenon that causes earthquakes like the ones in Plainfield: Fractured rock allows fluids to flow more easily. By its very nature, fracking is intended to cause earthquakes. The idea, though, is that they are supposed to be very, very tiny: microseismic, to be exact.

The USGS estimates that about a million wells were fracked between when Halliburton first revealed the practice and the end of the twentieth century. Most of them were traditional vertical wells. But as we continued to tap pockets of easily accessible oil, the industry had to look for new methods to find the less productive supplies, much of which were trapped in shale, a fine-grained, porous rock.

Shale tends to hold on to its oil and gas. To get them out, you have to really fracture the rock. These days, we depend upon shale for much of our gas and oil. And that means we do a lot of fracking.

As use of fracking technology increased, the industry began using an extraction method known as horizontal drilling. Instead of a traditional vertical well, which looks a whole lot like a borehole, horizontal drilling goes straight down until it hits a pocket of gas or oil and then turns. The resulting well ends up looking like the letter L. It's a highly effective practice for getting oil and gas out of shale in places like Oklahoma. It's even more effective when you use it alongside Halliburton's fracking technology. The industry refers to these wells as "unconventional oil and gas" (or UOG) development. At last count, there were over 100,000 of them in the country. Each one requires, on average, about 3 million gallons of fluid, mostly water along with chemical additives and sometimes an abrasive like sand, in order to break open the rock. For every three gallons of fracking fluid that go down, about one gallon comes back up.

Shale is also a great source for natural gas, of course. The combined process of horizontal drilling and fracking has now made the United States the world's largest producer of natural gas. Geologists are still somewhat split on the degree to which this process alone causes earthquakes. In the fall of 2016, a team of researchers at the University of Calgary published a catalog of seismic activity showing "that earthquakes are tightly clustered in space and time near hydraulic fracturing sites" in Western Canada. They also found a direct correlation between activity at these sites and when earthquakes occur. The official position of the USGS, however, remains the assertion that "fracking is NOT causing most of the induced earthquakes. Wastewater disposal is the primary cause of the recent increase in earthquakes in the central United States." Maybe, say the Calgary scholars. If it's true that fracking itself isn't causing the quakes in the United States, that could be because the geology is different from that in Western Canada. Then again, it could be that we just haven't proven the correlation between fracking and the quakes. They're not sure.

What we do know without a doubt is that moving fluid around underground redistributes stress, and that that alone can cause an earthquake. Scientists first observed this phenomenon in 1913, when a pair of Swiss towns that had never before witnessed an earthquake began to feel moderate rumbling. The national earthquake commission there eventually traced this freak shaking back to construction of a new tunnel. Apparently, workers there had accidentally bored through an underground aquifer and flooded the tunnel. Their gaffe had the immediate effect of lowering the town's water table, which was inconvenient enough. But then, a few months later, the earthquakes began. The commission concluded that redistributed water was to blame.

Back in the United States, wastewater injection remains a clear problem—particularly when it's returned to the ground in large amounts. In Oklahoma alone, that amounts to close to a billion barrels a year (or about 100,000 Olympic-size swimming pools). Some comes from fracking, but by no means all. That same brine valued by the Cherokee continues to define Oklahoma's oil fields today: Some of the less productive wells produce about a hundred barrels of brine for every one barrel of oil. In fact, the USGS estimates that just 10 percent of all wastewater injected into Oklahoma ground comes from fracking fluid.

Whatever its direct source, companies across the country have used different strategies to deal with all this water. In places like Pennsylvania, they've tried pawning off their fracking wastewater at public sewage plants. Others built huge wastewater pools aboveground. In places like Oklahoma, along with Ohio, Texas, and Arkansas, that water began going straight back into the ground—sometimes at rates as high as a million barrels a month. These injection or wastewater disposal wells now dot much of the Oklahoma countryside, particularly in places like Stillwater. They come with names like Deep Throat and Sweetheart and Jedi, as well as the names of old and current lovers: Joyce. Hannah. Linda. Earlier in the day, I'd asked a member of the Oklahoma Corporation Commission, which regulates oil and gas drilling, if Angela and I could visit one. He asked if the location would be off the record. I assured him it would be.

"Then I've got a big ugly one for you," he told me. The GPS coordinates arrived by text just a few minutes later.

It didn't take us long to get there: The well was located just south of the town proper. A sign advertised it only as a "commercial disposal" facility. A rusted red gate bore a standard-issue hardware store "No Trespassing" sign. Behind it, a gravel lane led to what looked like a big open-air pavilion. That's it, really. We stood around for just a few minutes, staring at all the nothingness, before a big rig appeared towing a bright red tank of wastewater. The driver pulled under the pavilion roof and opened the tank. Wastewater—about 10,000 gallons' worth—poured into a grate located at the center of a concrete pad. The whole process took under fifteen minutes.

As early as 2009, injection wells like these were receiving over 848 million barrels of wastewater a year. And as those levels increased, so, too, did the seismic activity. According to the USGS, the central United States historically saw about twenty-four earthquakes of a magnitude 3.0 or higher each year. In 2014, that number was 688. Most geologists say these quakes were probably caused by wastewater injection. Angela and the Crismons are more than sure. Mark shows me a map of the county. There are seven injection wells within a five-mile radius of his house and those two faults. Before the wells, he'd never felt an earthquake. Now they're a part of his daily life. In fact, as of this writing, Oklahoma is currently the most seismically active state in the Lower 48.

The quakes are getting bigger. In September 2016, an M 5.8 one ruptured just outside of Stillwater. Its waves shot out over six states and were strong enough that some residents in western Missouri were convinced it was the big New Madrid event. That 2016 quake, like most in recent history throughout the United States, occurred on a previously unknown fault. Unlike most other US quakes, it also occurred in close proximity to several big wastewater injection wells. State regulators shut down thirty-two such wells in the days after the quake. But that, says Spotts, kind of feels like too little, too late to her. It certainly doesn't fix the broken china and televisions, the toppled chimneys and shelves of groceries. In most cases, it's up to the owners of those things to clean up the mess.

Meanwhile, the earthquakes themselves are spreading out farther and farther. In November of 2014, a magnitude 4.9 quake rocked the town of Wichita, Kansas, breaking windows, creating chaos in homes and stores, and damaging at least two municipal buildings. Tremors were felt across 15,000 square miles. For most residents, it was their first experience with an earthquake. In fact, no one had ever even registered seismic activity there until January of that year, when pumping at two wastewater injection wells began to increase twofold. By July of that year, those wells were operating at five times the capacity they had been in January. And the earthquakes increased apace. The one in November was the second largest ever recorded in the state.

What precipitated it is visible even from space. Geophysicists at Arizona State University have begun studying data from a sophisticated Japanese satellite that uses a suite of radar and microwave sensors to chart the contours of the planet. What they've found is that the deformation created by wastewater injection is often visible in the months and even years before an induced quake. The rate of that uplift, coupled with information concerning ground pressures and pumping rates, could help communities predict these kinds of quakes before they occur.

The residents of Stillwater, Oklahoma, say they feel like their lives may depend upon that kind of technology. There's a huge fault zone called the Nemaha Ridge that begins just north of the town. It runs pretty much from Oklahoma City to Omaha, Nebraska. Scientists say, if it were triggered, it could produce a magnitude 7.0 earthquake. Maybe even higher. That's the kind of thing that keeps Mark up at night.

"We're in trouble," he says. "You activate that sucker and it'll kill a lot of people. And the problem is, you only die once."

Seismologists have been aware of the connection between fluid and earthquakes for decades, but it continues to surprise them. Paleogeologists and anthropologists have wondered for decades what prompted the collapse of Crete's Minoan civilization, located in an active seismic zone that includes the notorious Santorini volcano. During the nineteenth century, archeologists first found remnants of the civilization, which they believe was active from about 1700 to 1300 BCE. They were utterly wowed with the Cretans' technological savvy: Surviving palaces included storm drains, aqueducts and elaborate water mains, and what may well have been the planet's first flush toilet (true story). Rain and surface water provided some of the supply, but a lot of it came from a healthy series of wells and underground springs.

Sometime around 1500 BCE, Thera, the volcano that constitutes much of the Mediterranean island of Santorini, exploded in what may well be the most powerful eruption in our planet's history. It buried buildings in ash and single-handedly decimated the communities there. There is no evidence that ash reached as far as Crete. A British historian hypothesized that the island might have been wiped out by a tsunami prompted by the volcano, but wave models proved that would be next to impossible, and paleoseismologists found no evidence one had ever transpired. Many of the palaces on the island are still intact, which rules out a cataclysmic earthquake. There are plenty of cracks in the walls and foundations, but they were skillfully patched. Carbon dating showed these patches occurred over many years, which suggests that the inhabitants were accustomed to shaking.

So what did them in?

Right now, the best guess is that an earthquake robbed them of their water supply. Contemporary scientists think seismic activity might have been enough to drain the aquifer there and allow the inundation of seawater (today, the wells and springs there are far too brackish to be potable).

While this is maybe an extreme case, there are plenty of other precedents. The Chi-Chi earthquake radically changed water levels there. The Loma Prieta quake caused watersheds in Northern California to drop by as much as seventy feet in the weeks after the quake. The Hebgen Lake quake altered water tables across the country, including wells in Puerto Rico and Hawaii. Think about that. An earthquake in Montana—and one not even close to the strongest this country has experienced—was powerful enough to change water levels halfway across an ocean and a sea.

A few hours north of Pittsburgh sits the depressed town of Greenville. In 1998, back when a lot of the Main Street storefronts were still occupied, a 5.2 earthquake—the strongest in Pennsylvania history—ruptured nearby. I visited recently and stopped for lunch at Paxton's lunch counter. Connie, my waitress there, didn't have much to say about the historic quake at first. But after fried mush arrived and I began raving about this new-to-me culinary marvel (it really is exquisite—especially with maple syrup), she gave me the recipe. She also told me the quake had been strong enough to knock her daughter right out of her arms. She'd already lived through a tornado that leveled the town. But that quake scared her as much as anything.

It also had unexpected effects that still have hydrogeologists scratching their heads. In the hours and days after the quake, named Pymatuning after the eponymous reservoir nearby, a total of 120 households in the area lost their wells. Or, more exactly, they lost the water in their wells. All of it. Other homeowners turned on their faucets to find that their water looked like stale coffee and smelled like rotten eggs. Meanwhile, springs and lakes began appearing where there hadn't been any. Other wells overflowed and flooded backyards and basements. Gauges on the Little Shenango River rose. Mini geysers erupted in vegetable and flower gardens.

In time, some of the wells rebounded. Others stayed dry. The town is still one of the most studied hydrological areas on the East Coast. The USGS has a gauge at the Greenville Municipal Airport. It's a tiny spot up on a ridge with a turf and an asphalt runway, along with a gift shop that sells baseball caps and T-shirts with slogans like "What happens in the hangar stays in the hangar." The staff there are really accommodating. If you ask, they'll even show you the gauge on their way to get the afternoon mail.

～～～

In 1958, Charles Richter published his canonical work, *Elementary Seismology*. It's a huge book—nearly 800 pages—and dropping it could very well trigger an earthquake of its own (that's not actually true, but it could break your foot). By some accounts, it was the first definitive work on what was still very much a new science. In it, Richter addressed everything from the nature of ground motion to the effect of quakes on buildings and other infrastructure. He also addressed the different causes of earthquakes.

Richter wrote that he had been aware of the concept of induced seismicity after learning about a great quake in Asia; however, he all but dismissed the idea until they began occurring much closer to home: "The writer's earlier attitude toward such hypotheses has been modified somewhat by a remarkable series of disturbances under Terminal Island, in Long Beach Harbor, about 45 km from Pasadena," he wrote. "Perhaps another good instance of one being most impressed by events near home."

Those disturbances, he surmised, were directly related to extraction at the Wilmington oil field, a massive fossil deposit capable of producing anywhere between 3 and 6 billion barrels of oil. Drilling began there in the 1930s and soon became aggressive, with more than 3400 wells in operation, all withdrawing huge amounts of oil. As they did, the land on and around the field began to sink—as much as twenty feet in some places. And with that sinking, surmised Richter, came earthquakes. The largest was in 1949 and was strong enough to force 200 wells off-line.

Richter's hypothesis was confirmed in 2016, when USGS scientists Susan Hough and Morgan Page squared the historical catalog of earthquakes and oil extraction with what is known about contemporary induced quakes. They concluded that, in the case of these California quakes, it might actually be the *lack* of injected water that precipitated them: that removing oil caused the ground to subside, collapsing the rock around oil pockets and placing additional pressure on preexisting faults. They even went so far as to suggest that this practice may well have been the cause of the 1933 Long Beach earthquake, an M 6.4 quake that ruptured in the early evening on March 10, 1933. It killed 120 people and destroyed over seventy schools. Had the quake occurred when those schools were in session, the death toll would have been exponentially higher. Even now, the quake remains the deadliest in Southern California history.

Charles Richter also observed similarly induced quakes in Italy, which he saw as directly tied to the natural gas industry there, as well as in places like New York State, where salt extraction had prompted tremors. But the correlation remained murky until the army and a couple of chemical companies decided that injecting liquid waste deep underground was a good idea.

That story begins just outside Denver, Colorado, at the height of World War II. The previous world war had taught both sides a lot about the efficiency of chemical weaponry; by some counts, approximately 1.3 million casualties of the war were caused by chemical weapons, mostly in the form of gas. Both the 1919 Treaty of Versailles and the 1925 Geneva Protocol included language regarding the prohibition of chemical weapons. The United States nominally agreed to the idea, but Congress never ratified it and we made it known that we'd use them if we thought anyone else was. Meanwhile, we kept making them.

Much of that work occurred at the Rocky Mountain Arsenal, a 17,000-acre parcel of land northeast of Denver. As World War II raged, the US military used the facility to produce mustard gas and napalm, as well as white phosphorus, a particularly pernicious chemical that laces grenades and shells. Once ignited and in contact with human flesh, it burns to the bone. After the war, the army continued its chemical weapons project at the arsenal (mostly to make sarin gas) and also used the space to blend rocket fuel. Additionally, they leased sections of the facility to Shell Chemical for pesticide production.

All of these pursuits result in a lot of waste, and nearly all of that waste is highly toxic. The army tried burying it in trenches. They burned it. They allowed it to accumulate in storage pools. By the 1950s, residents nearby were complaining of contaminated drinking water. Cattle grew sick and died. So did thousands and thousands of migrating birds. Entire fields of crops languished. When state and federal agencies intervened, they found ground and drinking water contaminated with chlorine, phosphoric acid, arsenic, and more. One of the most troubling findings was the presence of 2,4-D, a powerful weed killer. No one could figure out that one, since 2,4-D wasn't even *produced* at the plant. Eventually, scientists reached a disturbing conclusion: While the herbicide wasn't formally made by humans at the arsenal, all of the ingredients were there. Somehow, as it coursed through groundwater and streams, it had made itself.

The army decided that the only way to get all of this toxic material out of the environment was to inject it far below the water table. And so they began to dig one of the nation's deepest wells. They didn't stop until they had gone down over 12,000 feet. At that depth, their geologists insisted, the rock was "capable of accepting fluids injected under pressure." In March of 1962, they began testing that theory, injecting 43 million gallons of waste at rates of over two gallons per second.

The first earthquakes began shaking the region in June of that year. Four occurred on the evening of June 18. They weren't big enough to do much damage, but they did slam around the doors of the Adams County jail, leading sheriff's deputies there to assume a noisy escape was under way. As the year progressed, so did the earthquakes. And they got stronger. When interviewed by UPI, an international news agency, in December of that year, Dr. George Meredith of the Colorado School of Mines told reporters, "It is more hazardous to predict earthquakes than it is the weather, but it appears there could be more."

He was right. The grand total was 189 quakes by the end of 1962. And with the start of the New Year, the quakes continued to increase in both intensity and rate—almost doubling the number of the previous year. The strongest in 1963 was over a magnitude 4.0. It occurred during the wee hours of the morning and jolted Denver residents out of their beds. One woman said it felt like her "house was turning over."

The overwhelming majority of these earthquakes had an epicenter within five miles of the arsenal injection well. And still, no one drew the correlation between the arsenal and this unstable ground. As far as I can tell, it's because very, very few people knew the injection was happening. I scanned every public document and newspaper I could find for 1961 and 1962. The Rocky Mountain Arsenal was mentioned plenty of times: They had been chosen by the Pentagon to create a "new war gas," their basketball team was selected for the Fifth Army Basketball Tournament, their deer were seen migrating en masse to a nearby airfield, but there is no mention that I can find of the injection project. In fact, when asked by the *LA Times* in June of 1962 about the quakes, officials there "were unable to explain what happened."

The first person to suggest a correlation was Father Joseph Downey, director of the Regis College Seismological Observatory. He was soon joined by a consulting geologist, who got the attention of a local congressperson. That official, in turn, was able to leverage a full investigation. In 1966, the USGS concluded that the earthquake swarms were, in fact, produced by the arsenal well. In seismological circles, the incident has since become the first—and one of the most irrefutable—cases of induced seismicity not just in the United States but also in the world.

Injection ceased in February 1966. But the earthquakes continued. Their epicenters began moving out and away from the arsenal itself. And they got a lot bigger. In April of 1967, the strongest swarm yet shook the region. The tremors broke 118 windows at the arsenal. They damaged schools as far away as Boulder. They stopped congressional proceedings at the statehouse. By the end of the year, the arsenal had produced over 1500 discrete seismic events.

Army geologists had to face a terrifying fact: While they clearly had the capacity to create the quakes, they appeared powerless to stop them. And while the shaking did decrease over time, the tremors continued to plague people in the Denver area for nearly twenty years.

You're welcome to visit this seat of American induced seismicity. In fact, the US Fish and Wildlife Service would love it if you did. The area was declared a "national priority superfund site" in 1987, just after the last of the shaking subsided. That didn't stop local wildlife from moving in: coyote packs, deer, and a whole host of birds, including the then-endangered bald eagle. As they watched this rewilding unfold, the USFWS proposed the idea of a wildlife refuge. They converted one of the remaining arsenal buildings into the National Eagle Repository, which serves as a clearinghouse for "the receipt, storage and distribution of bald and golden eagles found dead and their parts throughout the United States." The eagles, and their parts, are then in turn issued to Native Americans and Alaskan Natives enrolled in federally recognized tribes for use in Indian religious ceremonies.

Meanwhile, the army and the EPA agreed upon a remediation project to clean up their mess. It mostly included clearing out contaminated soil down to ten feet and covering landfills so that humans and animals can't access them. They left some of the weirder parts of the arsenal, like the cement stops in parking lots and menacing signs warning you not to approach the signs. The rest they blanketed in landscaping fabric and let it grow over. There's a gleaming new visitor center as well. I stopped in with a friend just before they were closing. She'd recently moved to the area and was worried about what induced seismicity might mean for her six-year-old daughter. We took in the carefully curated displays, most of which touted the success of the restoration project. Tucked behind them was a much smaller interpretive center dedicated to the munitions projects. "Boogie Woogie Bugle Boy" cranked over the loudspeakers there, and most of the signage promotes the incredibly patriotic contributions of the arsenal (along with its award-winning bowling alley—seriously). One placard covers the earthquakes. Whoever authored it is far less persuaded of the causality than most contemporary geologists. "Deep well injection had been safely used elsewhere as a water disposal method for many years," it insists (somewhat defensively, if you ask me). "Some experts believed that injecting fluid into the Arsenal's underground well caused the tremors." Below it is an article from *The Denver Post*, dated November 1965: MORE EVIDENCE WANTED, reads the headline.

We asked the USFWS naturalist docent on duty that day about the seismic history of the place. She seemed visibly wary of my notebook.

"We try not to focus on what the land was," she eventually offered. "Instead, we focus on what it is becoming."

THE EARTHQUAKE FIGHTERS
OF OKLAHOMA

idwesterners really, really love football. The only region that
might love it more is the South. Oklahoma arguably has a foot
in each region. At 86,000, the capacity of the University of Oklahoma Sooners stadium ranks about fifteenth on the list of the country's biggest college stadiums. That's plenty big to create seismic
activity: Seattle's CenturyLink Field, home of the Seahawks, holds
just 67,000 people and regularly makes an appearance on the Pacific Northwest Seismic Network monitoring stations. The biggest
event, reports CNN, occurred in 2013 when a defensive lineman
grabbed a fumbled ball, ran twenty-two yards, and scored a touch-
down. The crowd went nuts, and so did the regional seismographs.
After realizing this was a regular occurrence, network staff installed seismographs in the stadium proper, hoping to monitor not
just how built structures respond to seismic activity but also how
crowd-sourced information might help with early warning systems.
It also gave a group of geologists an opportunity to watch a couple

of downs and maybe grab a hot dog or two. Their blog explained it this way: "While nowhere near as stressful and dangerous as a real geophysical event, a Seahawks game can allow us to test and practice many of the capabilities we will need sometime in the future . . . and will be fun, besides."

They're not alone. Seismologists have recorded similar activity at British soccer games and after Louisiana State University came from behind to beat Auburn in 1988 (that game and its resulting seismic activity earned a spot at one of Ripley's Believe It or Not odditoriums. So it seemed perfectly natural to Katie Keranen, a seismology professor at the University of Oklahoma, to install a series of seismographs at the Gaylord Family–Oklahoma Memorial Stadium the first weekend in November of 2011. The Sooners were playing Texas A&M that Saturday, and attendance was predicted to be high.

Kickoff was scheduled for 2:30 P.M. Both ABC and ESPN were broadcasting the game. Keranen's students never made it. In the wee hours of the morning, an M 4.8 quake rocked the state, with an epicenter located in the tiny town of Prague, just over sixty miles to the northeast of the university. As soon as they learned of the quake, Keranen and her students grabbed their equipment and made the drive. The students had just reinstalled their gear when another earthquake—this one much larger, at M 5.6—rocked the town. Over a half million people residing in seventeen different states reported feeling the quake, including as far away as Wisconsin and Tennessee.

One diner owner likened the experience to being on the high seas: "It rolled kind of just like a boat or a ship would roll underneath your feet," he told *The New York Times*. For close to a minute, it seemed to lift parts of his diner, sending broken plates everywhere. Fifty miles away, in Oklahoma City, a group of ghost hunters had set up video cameras at a Mexican restaurant purported to be haunted.

They had very much the same experience. "The video looks like it was recorded on a boat in rough seas," one reported. Their footage is available on YouTube. I've been on a boat in rough seas, and I have to say their videos seem a lot more like what I'd imagine paranormal activity to look like. In one, a back hallway of the building bends like a fun house. In another, the video camera does a 360 not unlike Linda Blair's head in that scene in *The Exorcist*.

By the time that second quake struck, Keranen was back at her own house in Norman. The shaking there lasted twenty seconds. Keranen knew it was a big event. So she rounded up more monitoring devices and as many students as she could muster on a Sunday morning and headed back out into the field. As far as she could tell, the quakes and their aftershocks were centered on the Wilzetta oil field, part of a larger fault system by the same name. The area—and the faults—had been a by-product of the formation of the Rocky Mountains some 300 million years ago. Oil drilling began there in the 1950s, and fluid injection commenced sometime after 1993. The wastewater wells had operated there for twenty years but had dramatically increased their injection rates in recent years.

Meanwhile, a similar earthquake had rocked the town of Youngstown, Ohio, and sent palpable waves as far away as the East Coast. It was the most recent of eleven quakes to shake the area that year. Epicenters for nearly all of them were located within a mile radius of the Northstar 1 deep disposal well. In December 2010, the well had begun injecting wastewater over 9000 feet into the basement rock. The first tremors began three months later. State officials asked the owners of the well to cap it. They agreed, but said they wanted to do more pumping first. The plan was to inject the water into sandstone not too far below the surface. But the sandstone turned out to be not nearly as porous as the well owners had hoped, and it took a lot of pressure to get the water belowground.

Worried that they could not determine just how much water was making its way through the basement rock, state officials asked scientists at Columbia University's Lamont-Doherty Earth Observatory to investigate. They deployed their devices on December 1. On Christmas Eve, they recorded an M 2.7 quake in the area. Four days later, they presented their findings to state officials. The Ohio Department of Natural Resources ordered the well to be shut down and were there to observe that it cease operating at 5:00 P.M. on December 30. Barely twenty-four hours later, the biggest earthquake yet—M 4.0—surprised New Year's revelers as far away as Salem, Massachusetts.

Ohio state geologists were able to conclude that water from the injection well had triggered a previously unmapped fault that ran under the city. The governor agreed and ordered a moratorium on three similar wells.

Oklahoma officials saw it differently. As the aftershocks continued to rumble (over 1000 in all), Katie Keranen began work as the lead author on a study that would eventually be published in 2013. The trick was to prove the connection between the wells and the earthquakes. For that, she says, seismologists must demonstrate three things: a close proximity between wastewater wells and earthquake epicenters, a demonstrable increase in the area's seismicity, and a connection between the event and parameters with which the wastewater was being injected. She had all three.

The state of Oklahoma disagreed. In a statement issued after the release of Keranen's study, then state geologist Austin Holland wrote, "It is still the opinion of those at the Oklahoma Geological Survey that these earthquakes could be naturally occurring. There remain many open questions, and more scientific investigations are underway on this sequence of earthquakes and many others within the state of Oklahoma." The Oklahoma Geological Survey had also been looking at the data from the Prague quakes. Based on their analysis, "the interpretation that best fits current data is that the Prague Earthquake Sequence was the result of natural causes."

In the days following the Prague quake, the governor of Oklahoma, Mary Fallin, tried to avoid the subject as much as possible. She was preparing to speak at an oil and gas trade conference. When the marketing firm coordinating events asked if she would be making a statement about the recent quake, her communications director declined: "That's probably actually not a great topic," he wrote in emails obtained through the Oklahoma Open Records Act request by *Energywire*: "She could certainly say, 'yeah that was crazy.' The problem is, some people are trying to blame hydraulic fracking (a necessary process for extracting natural gas) for causing earthquakes. This is an energy conference heralding natural gas as the energy source of the future . . . so you see the awkward position that puts us in."

Instead, he sent a series of "potential talking points." Included among them were: "There is no current evidence that oil & gas operations had anything to do with the recent large earthquakes in Oklahoma" and "According to the OGS, the earthquake characteristics of both intensity and depth essentially rule out man-made causes." For her part, Katie Keranen says she experienced a similar response from administrators at OU. She ended up leaving the school to take a position at Cornell.

That wasn't an option for people like Angela Spotts and the Crismons. Besides, Spotts says, she's always been an activist. She drives a Prius emblazoned with liberal bumper stickers advocating for peace and no more corrupt government and an end to climate change. She wears T-shirts declaring, "I'm an OK earthquake fighter," and is perfectly comfortable behind a megaphone or underneath a handmade protest sign.

Mark Crismon couldn't be more different. He sneers at the concept of activism, saying it's some kind of "hippie leftist feedlot graffiti." That's his polite shorthand for "bullshit." Mark thinks a lot having to do with the earthquakes and the state's response is feedlot graffiti.

"Don't call me an activist," he scolds me when I ask. "I deal in reality. Activists are tree huggers. I'm not a tree hugger in any way, shape, or form. That's a left-wing thing." Despite their different social views, he and Angela have nevertheless formed an unlikely alliance because of this issue.

Spotts is the cofounder of Stop Fracking Payne County. Mark is their information guy. He's got the seismometer data, the war room, the longtime local contacts. But they needed more. They needed expertise. And so they invited Oklahoma State University professor Todd Halihan to speak on the issue at the local library. They expected a couple of dozen people. Halihan packed the place.

Halihan is tall, gregarious, even goofy. He laughs a lot. His sense of humor is built squarely on a fourteen-year-old's sense of irony. I asked if we could pal around together for a couple of days at the end of his spring semester. That first evening, I attended one of his hydrology courses. I figured I'd blend in if I wore cowboy boots. His students didn't fall for it. When they asked what the deal was, Halihan told them I was a random homeless person who started following him earlier in the day. Earlier that academic year, when Hanukkah came around, he built a menorah for a Jewish colleague—made out of canisters of Spam. He'll tell a room of devoted Baptists worried about earthquakes that they're all screwed. In the wrong hands, this kind of humor could seem too shock jock. But Halihan is exceedingly kind and compassionate and so obviously full of Midwestern conscientiousness that it somehow just works.

Halihan was one of the first scientists in Oklahoma to speak out publicly about the connection between wastewater injection and earthquakes. He didn't set out to become a celebrity. He certainly didn't want to become a crusader. He just wanted the quakes to stop.

"I have a son," he says. "I worry about him. I can't bear the idea of someone pulling him out of a collapsed school building."

So he gave a talk. And then another. Mostly, he says, he just rehearsed the known science. Sometimes, he read from peer-reviewed literature. "You know," he says, "like story time."

In no time, he was receiving dozens of requests from around the state. He tried to say yes to all of them. He was careful never to do it on OSU time—or their nickel. He wanted to stay neutral.

In his evening class, I sit next to a young woman clad in tight jeans and a baggy T-shirt. She's slender, she wears perfectly applied eyeliner, and her long dark hair is pulled into a tight ponytail. An open Mountain Dew bottle and a Slim Jim sit in front of her. She explains to me that, by day, she's a mud logger: She's employed by one of the big oil companies to spend eight or even ten hours a day in a tiny portable trailer analyzing drill samples. Sometimes, she's installed in "man camps," where workers from across the country and mostly the Deep South work twenty-four hours a day drilling for oil and gas while living in packed mobile homes. Another student tells me her family has had a pumpjack on their property for decades. They recently agreed to a disposal well.

In other words, the situation is complicated.

That's true for the state as well. The fortunes of Oklahoma have long since hung on oil and gas. It's a state populated by billionaire oil tycoons and families struggling to get by. That's particularly true in places like Payne County, which has a poverty rate of over 24 percent. A lot of times, energy companies fill the gap. That alone isn't all that unusual: Seven states in the United States derive more than 8 percent of their GDP and 5 percent of their total personal income from oil and gas. Of these, Oklahoma is in the top three when it comes to dependence on the industry, particularly where jobs and personal income are concerned (the other two are North Dakota and Wyoming). And yet Oklahoma consistently comes in last among these states where things like average home price appreciation is concerned. One reason why is because other states, like

Texas, have done a much better job of diversifying their industries to include things like financial services, manufacturing, and transportation. Right now, about 25 percent of all Oklahoma jobs are directly tied to oil and gas. The companies are literally everywhere in Oklahoma. They donate to schools and create community centers. Their proceeds build municipal swimming pools. They also finance election campaigns and, some would argue, the statehouse itself. A pumpjack sits outside the capitol building, where it once drew from one of Oklahoma's most lucrative oil fields.

Higher education there is no less complex. The legendary oil wildcatter and corporate raider T. Boone Pickens has donated over $500 million to Oklahoma State University. The school of geology bears his name. The labs there are endowed by energy companies. Until 2013, the president of the university served on the board of Chesapeake Energy, the second-largest producer of natural gas in the country. Many of Halihan's students work in the industry—or aspire to. Angela and the Stop Fracking Payne County group started putting up signs around the area, saying they wanted the quakes to stop. One morning, she woke up to a neighbor running over her signs with his pickup truck. Other signs around the town began disappearing. So she started locking them in place with bike locks. The locks got cut.

When Halihan started giving his talks, he worried about their impact on his job. He says he did receive some pointed suggestions from members of the administration that he remain neutral. "But," he says, "there were others who were saying, 'Keep going.'" So he did. He's received his fair share of criticism, but he says it mostly seems to balance itself out. "I've been accused of being an activist, of working for the industry, of being too neutral."

Life in the spotlight has come with a steep learning curve—like when a Korean news crew teased him for showing up in a plaid shirt (Korea invests heavily in US fracking, and prints look wonky on TV). Over time, he's developed an ever-growing list of rules for doing this kind of work. Some of them are simple, like "Always be on time and stay to the end." It's important, he says, to make sure that concerned people feel heard. Others are more complicated, like understanding what "expert" means to politicians and to other scientists (it's not necessarily the same thing).

He says there have been times when he's considered standing up and saying that he doesn't want his son to die. But he worries that kind of emotional appeal might backfire. A colleague of his, a sociologist, recently gave an impassioned talk about the effects of these quakes on the community. She was in tears by the end. She and Todd are good friends, and he's empathetic, but he also had some stern words for her afterward.

"This is like that movie *A League of Their Own*," he explains. "There is no crying in baseball."

That may be one reason why Mark and Angela see Todd Halihan as their not-so-secret weapon. In fact, Mark won't even talk to Todd, for fear it might taint his effectiveness. They've never even met in person. But Stillwater is a small town. They know enough about each other.

Halihan also has a seismometer. When I first met him, it was stationed in his backyard, near a pond overflowing with turtles. His son and I fed them one evening. We'd scatter what looked dog kibble over the surface, then wait for the reptiles to arrive. They did—seemingly by the dozens. Later, we sat with the rest of the family over grilled burgers and talked about what life is like for them. Todd's mother-in-law lives with them. She helps Todd and his wife take care of their son. They're all equally concerned about the earthquakes.

"They feel like an eighteen-wheeler hitting the side of your house," said Todd's mother-in-law.

Halihan agreed. Still, he said he felt like job security dictated that he pick his battles carefully.

A year later, I call to check in. When I do, Halihan says he no longer worries so much about what impact his public talks might have on his job. The scholarship on wastewater-injection-induced earthquakes is mounting. The USGS acknowledges it as a real threat.

"It's like climate change," he says. "You can deny the science, but at this point, the evidence is irrefutable."

In 2016, OSU gave him an award honoring his public outreach (though they spelled his name wrong on the plaque). Halihan didn't correct them. And he's hopeful about some of the changes he's seen in the state. In 2015, the Oklahoma Geological Survey reversed its stand on the quakes, issuing a statement that read, "Based on observed seismicity rates and geographical trends following major oil and gas plays with large amounts of produced water, the rates and trends in seismicity are very unlikely to represent a naturally occurring process."

The Oklahoma Corporation Commission now has unilateral power to shut down problematic wells. But the efficacy of this power may be diminished over time. It costs about a quarter of a million dollars to plug a problem well. In 2015, the OCC ordered 140 plugs. That cost the industry upward of $40 million.

And the more water you inject, the harder it is to isolate which well is causing the problem. At the Rocky Mountain Arsenal, earthquake epicenters migrated over six miles away from the injection site. In 1991, the US Bureau of Reclamation began pumping much of Colorado's Dolores River underground. (The river, a tributary of the Colorado, ran over salt beds, which rendered the flow unusable for irrigation. The idea was to bury the water below the salt and preserve the Colorado River.) Since that project began, the bureau has recorded more than 5700 seismic events. Many are too small to be felt, but an M 3.9 was sufficient to shut down the plant, and the highest topped out at M 4.3. The longer the project operated, the farther away the resulting earthquakes migrated. Today, hydrologists aren't sure just how far injected water can seep, but they have evidence the resulting quakes can migrate as far as fifteen miles from their source. In Oklahoma, a state where it's not uncommon to have multiple injection wells within a square mile or two, it is getting harder and harder to link an earthquake to a particular well.

Meanwhile, the earthquakes around the Halihan house continue. It still feels like an eighteen-wheeler runs into their garage, sometimes a couple of times a day. But now the house also gets this weird vibration: "Like you're living in a back massager," Todd explains. His seismology colleagues have speculated that that might have something to do with the way P and S waves emanate in an induced quake (seismologists still aren't sure to what degree these induced tremors resemble natural ones). And so Todd and the OSU researchers have moved his seismometer into a spare bedroom directly above the living room where he and I first met. It's sensitive enough to register when his mother-in-law gets up to use the bathroom in the middle of the night. The jury's still out on what it can tell Halihan about the induced quakes below his house.

TANK FARMS ON THE PRAIRIE

Induced earthquakes have dramatically changed the face of seismicity in this country. The USGS hazard maps have struggled to keep apace. The Prague quake, for example, was "significantly stronger" than anything forecasted in that area by USGS hazard models. Historically, says Art Frankel, the USGS has always omitted induced quakes from their calculations: Those tremors just seemed too freakish, too minor to be part of a sustained study. The Prague earthquake was a game changer. In 2014, the USGS convened a panel of experts from universities, governmental agencies, and private firms to address the issue of induced seismicity. They decided it was time to include this phenomenon.

Mark Petersen was the lead author of the resulting report. He says, in the end, the decision came down to public safety. "We want to help people understand how much concern they should have with these earthquakes," he said. "By including human-induced events, our assessment of earthquake hazards has significantly increased in parts of the US. This research also shows that much more of the nation faces a significant chance of having damaging earthquakes over the next year, whether natural or human-induced."

Still, it's hard to figure out how to account for the hazard these quakes pose. To varying degrees, both of those considerations become real sticking points when trying to forecast the likelihood of future induced quakes. As we've seen again and again, humans have the ability to change geology, and that can also change seismic risk. It doesn't matter if there's no record of previous shaking in the area. Induced earthquakes are highly dependent upon decisions made by humans about whether to drill or inject fluid and for how long. A computer model that anticipates these decisions doesn't yet exist. Whether drilling or wastewater injection is going to cause an earthquake in a particular place is still very much a question as well.

Take the Bakken oil formation, a huge deposit of oil and natural gas that reaches from North Dakota and Wyoming up into Manitoba and Saskatchewan. North Dakota alone has over 400 wastewater injection wells, and they've seen over 30 million barrels of water since they began operating. However, the area's seismicity has barely changed. Scientists aren't entirely sure why: They think it could be because the area lacks heavily loaded faults, or that maybe the amount of water injected hasn't yet hit a critical level. Pennsylvania has injection wells, too, but as far as we know, none of them have sparked an earthquake. Hydrologists for the EPA think that's because the water there hasn't yet penetrated the basement rock. It did in Arkansas. There, in September 2010, a continuous swarm of earthquakes began along a previously unknown fault between the towns of Greenbrier and Guy in the north-central portion of the state. Over the next six months, residents there endured about 1000 quakes, the strongest measuring M 4.7—enough to crack homes in two. The state responded by imposing a moratorium on injection in the area that included a mandate to plug existing wells and a prohibition against any new ones. The decision bankrupted Clarita Operating, a small company that ran several of the wells.

The capricious nature of when wells are active makes it hard to forecast hazards to a particular area over any amount of time. Initially, to figure out what to include, hazard specialists at the USGS combed through scientific literature, trying to create both a timeline and a map of quakes believed to be induced. They tried to keep abreast of what communities had put a moratorium on processes like fracking and wastewater injection, along with other attempts to manage the processes. In the end, they decided that the amount of variability really only allowed them to project hazard rates for a year or two.

They're still scratching their heads about the intensity of induced quakes. Some scientists and lawmakers say that the Prague, Oklahoma, quake is as big as we're going to see. Others point to a series of earthquakes centered on an Uzbekistan natural gas field. Those topped out at M 7.3. In China, an M 7.9 quake is widely believed to have been prompted by a reservoir there. There's plenty of cause, then, to believe we could see a bigger induced quake as well.

Seismologists also worry about a phenomenon known as remote triggering. Ross Stein, a geophysicist with the USGS Earthquake Hazards Team, was one of the first to suggest the idea that earthquakes engage in "conversations" with one another, particularly on a single fault system. The basis for his hypothesis sprang from a peculiar series of earthquakes in June of 1992. The first, a 7.3 quake, occurred near the city of Landers, California. Just three hours later, a 6.5 quake ruptured twenty-five miles away in the town of Big Bear. Not long after, an M 5.6 quake rocked Nevada's Little Skull Mountain, some 150 miles away.

The quakes, while obviously near one another, were too far apart to be considered fore- or aftershocks from the main event. Ross hypothesized that the stress of the first quake, rather than fully dissipating, had instead traveled down the fault and ruptured at another loaded site. A few years later, he found further evidence on Turkey's North Anatolian Fault. There, scientists analyzed twelve quakes that had occurred on the fault since 1939. And they used that data to successfully predict a large event near the city of Izmit. It ruptured in 1999, causing $6.5 billion in damage and killing approximately 25,000 people. Just a few months later, another quake ruptured even farther down the fault. A review of all the major quakes in the world over the span of about twenty-five years found that the majority of them occurred in sites where an earlier quake had increased the stress.

So how far can this seismic stress travel in order to trigger another quake? That's one of several million-dollar questions in seismology right now. Heather Savage believes an M 4.1 Oklahoma quake could have been triggered by a massive quake in Chile just sixteen hours prior. She and her colleagues published a paper arguing that these megaquakes can send ripples of seismic waves strong enough to trigger events at wastewater wells around the world. There's evidence that the massive 8.6 quake off the coast of Sumatra may have prompted two significant quakes in Japan and another near Nenana, Alaska. In the case of the Japanese quakes, seismologists there theorize that a 2011 event there had loaded faults to the point that they were easily triggered by the Sumatran event. Bob Smith is convinced that both the 1964 and 2002 Alaska quakes triggered seismic activity in Yellowstone—over 2000 miles from the Alaska epicenters. The latter one, an M 7.9, prompted a swarm of over 250 quakes in the Wyoming park.

We're also learning that earthquakes can hop around a zone. Prior modeling always assumed that ruptures were confined to a single fault or that only single segments of a long fault would rupture. Several recent earthquakes have proven otherwise.

This is what really worries residents in the Midwest. Earthquakes like the ones in Prague are scary enough. But there are some big faults in Oklahoma, including the Nemaha Ridge. If that were to be triggered, says Halihan, we'd be talking not about how few people died but, rather, about how few lived.

The USGS estimates that approximately 7 million people live in areas that have seen a notable increase in hazards created by induced earthquakes. That includes significant population centers like Dallas and Oklahoma City. Looking at the new hazard maps, there's no denying it: A big red bull's-eye surrounds the area.

In Oklahoma, the number of stronger earthquakes, which is to say earthquakes with a magnitude of 3.0 or higher, continues to increase. In February 2016, the state had an M 5.1 quake, strong enough to damage schools and worry some officials. That quake was just seventy-five miles west of Cushing. Nine months later, an M 5.0 quake struck the town proper. In a lot of ways, Cushing is just your average struggling town. In a lot of ways, it's not.

Back on March 12, 1912, another wildcatter—this one by the name of Mad Tom Slick—struck pay dirt on the Wheeler farm, just outside of Cushing. That afternoon, over 200 barrels geysered out of the ground before workers could cap the hole with an old metal washtub. In the days and weeks that followed, the Wheeler farm became a work camp—a sea of makeshift tents housing hopeful men from throughout the region. And the oil kept coming. By 1915, over 700 wells on Wheeler's fields were producing 1.2 million barrels of oil. Each day, pipelines flushed 40,000 of those barrels to refineries in Texas and Kansas; 200 railcars carried the crude to additional markets every day, too. But that wasn't nearly enough. And so the oil spilled onto the land, pooling in low places and contaminating fields. Companies built refineries by the dozens in an attempt to keep up with production—in 1920, there were fifty of them. Thirty gasoline plants toiled nearby. But still it wasn't enough. So companies began building enormous steel tanks to hold the oil until it could be moved or processed. More workers poured in to help with construction. Towns with names like Capper, Drumright, and Oilton began to pop up between the tank farms. They brought with them demand for opera houses and creameries, brothels and taverns.

As it turned out, production in Cushing had reached its peak by 1916 and slowed every year following. First refineries shut down, then stores and theaters. Then the railroad stopped service to Cushing. Towns began to evaporate. But the tank farms remained and grew—fed by an increasing number of pipelines leading to and from this small Oklahoma town.

These days, the name of the oil game in Cushing is light sweet crude. Like maple syrup, all oil gets a grade: from so light it can pretty much power your car right out of the ground, to a heavy so viscous you can drive that car on top of it. Oil is also designated either sweet or sour depending upon its sulfur content. Why sweet and sour? Because a hundred years ago, oil companies staffed professional tasters who swirled and quaffed it like twenty-first-century sommeliers.

Of all the combinations of oil viscosity and taste, light sweet crude is one of the most valuable, since it converts to gasoline with little waste or refinement. In 1983, the New York Mercantile Exchange made Cushing their clearinghouse for the stuff. The town of 7000 was still dirt-poor, but now it was the epicenter of the oil world—to a tune of about $179 billion annually. More and more pipelines snaked into the region—the Pony Express from Wyoming, Spearhead from Illinois, Seaway from the Gulf Coast. The Keystone Pipeline arrived later, bringing with it over 500,000 barrels a day. The tank farms grew. And grew. And grew until they could hold more than 80 million barrels—more oil than you'll find in any other place on the planet. There are about 300 tanks around Cushing now. Some of the larger ones can hold 575,000 barrels of oil. Collectively, they are the single largest concentration of oil on the planet. Six major oil companies own and control the tanks, but dozens of others lease space for their oil. And while the pipelines are always moving oil, they also represent a real international nexus: Everyone driving into town is greeted by an enormous pipeline sculpture bearing the town's motto, "Pipeline Crossroads of the World." That's no mere PR bluster: Fourteen major ones cross through Cushing, making it North America's biggest oil hub—and quite possibly the world's.

This concentration of oil and companies and infrastructure

makes Cushing a fiercely competitive market for oil, with the actions of one player capable of exerting real influence on all the others and, ultimately, on us as consumers, too.

The Department of Homeland Security has designated Cushing "critical infrastructure," which they define as assets that, "whether physical or virtual, are considered so vital to the United States that their incapacitation or destruction would have a debilitating effect on security, national economic security, national public health or safety, or any combination thereof." In a 2013 report on energy infrastructure, the department acknowledges that natural disasters pose the greatest threat to that infrastructure, though they mention neither Cushing nor earthquakes specifically.

The tanks there are built to withstand tornadoes. No one really knows to what degree they can withstand an earthquake. A 2015 study published by USGS and university geologists concluded that the Wilzetta-Whitetail Fault Zone just south of Cushing is capable of producing "very strong shaking levels" capable of "causing moderate to heavy damage" to the pipelines and oil tanks there. I called each of the six major companies that own these tanks, asking for information about what seismic protocols are in place for them. I repeatedly emailed, requesting visits or interviews with safety officers. Not a single one returned my calls.

Tom Heaton at Caltech says most, if not all, of the tanks in Cushing are built to the weakest design standards established by the American Petroleum Institute, and that those standards do not take induced seismicity into consideration. He thinks a quake like the one in Prague could be enough to violently push the oil from one side of the tank to another. In geological terms, the phenomenon is known as a seiche: an internal wave or oscillation of a body of water. It's the same thing that made settlers think rivers were running backwards during the New Madrid events and what caused the University of Arizona's swimming pool to lose water after the 1985 Mexico earthquake (which was over 1200 miles away!).

Just how much damage is enough to affect the average American consumer?

Ron Ripple is the Mervin Bovaird Professor of Energy Business and Finance at the University of Tulsa. He's been studying the business of oil for decades. He cautions that it's hard to quantify hypothetical scenarios like these, but he estimates that an earthquake or other disaster would have to knock out half those tanks to have a real impact on the market. Of bigger concern to him are the pipelines, which control a larger volume of oil. He points to the October 2016 explosion of the Colonial Pipeline in rural Alabama as a corollary. The resulting fire kept crews from repairing the pipeline proper for six days. During that time, oil commodity prices jumped 60 percent—the highest spike in nearly a decade. Exporters clamored to find work-arounds, including tankers capable of moving the oil by sea. As a consequence, freight cargo rates increased by nearly 40 percent. Meanwhile, motorists in southern states rushed the pumps, elevating prices there, too—forcing the governor of Georgia to issue an executive order warning about price gouging.

It wouldn't be unreasonable, says Ripple, to see a similar scenario were the Cushing pipelines to go down. The Colonial Pipeline moves about 100 million gallons of oil and gasoline a day—about the equivalent of the Seaway Pipeline, just one of the more than a dozen that converge on this town. That pipeline was also shut down in late 2016, after authorities in Cushing noticed a spill. The effect of that shutdown had the opposite effect, pushing the price of US oil below fifty dollars a barrel, as international traders worried they wouldn't get their deliveries.

"Prices move through the markets fairly quickly," says Ripple. "We tend to see opportunistic changes in prices right after an event. Some of those look like a pretty close cause-and-effect relationship between supply and demand. Other times, you'll see impacts that leave us all scratching our heads. In the end, you just don't know how the market and consumers will react."

Johnson Bridgwater, director of the Oklahoma Chapter of the

Sierra Club, says he's mindful of the economic effects of such a spill, but it's the impact on the landscape and the people who occupy it that most concerns him. In 2014, he and his organization began traveling around the state, holding public meetings about the danger of quakes. Oftentimes, they brought Todd Halihan along with them.

"We always want our conversations to be rooted in objective science," Bridgwater explains. "This should be about fact, not emotion."

Nevertheless, as the two traversed the state, they heard a lot of concerns from communities and the individuals who populate them. And as the earthquakes grew in size and number, so, too, did the concerns. Bridgwater says it felt like the state wasn't listening to them. And so, in 2015, his organization filed a federal suit against three of the largest natural gas companies operating in Oklahoma—Chesapeake, New Dominion, and Devon Energy—under the Resource Conservation and Recovery Act (RCRA). Originally known as the Solid Waste Disposal Act, the act was passed in 1965 to empower Americans and compel polluters to take responsibility for their waste. It was amended in 1970 to broaden the scope and to give the federal government more jurisdiction over its enforcement. Historically, suits filed to enforce the RCRA have been attempts to force cleanup of landfills, stop the illegal dumping of hazardous waste, or seek remediation of harmful industrial practices. The Sierra Club's suit, says Bridgwater, is a real test of the act and the reach of federal oversight.

"The act has never been used in this way," says Bridgwater. "But then again, we've never had a state experience twenty thousand earthquakes, either. The whole situation is unique."

Specifically, the Sierra Club is asking for significant reductions in wastewater injection volume in and around Cushing. They're asking for an independent scientific body to oversee earthquake response, and in the meantime, they're also asking that the named defendants reinforce their tanks, pipelines, and other vulnerable infrastructure.

It's a twofold problem in places like Oklahoma, says Bridgwater.

The actual water being injected, he points out, contains 260 toxic chemicals. In 2015, the EPA reported that approximately 6800 sources of public drinking water serving nearly 9 million people exist within a mile of wastewater injection wells. That same study referenced 225 reported spills at wastewater injection sites; its authors weren't able to tally the number of spills at actual fracking sites, though, in both cases, the most common reason for a spill was a "failure of container integrity" at the site. In North Dakota, a single broken pipeline spilled 2.9 million gallons of wastewater, contaminating both surface water and groundwater in the area. Bridgwater worries an earthquake could allow similarly toxic wastewater to infiltrate the Oklahoma water supply.

And then there is the oil itself. The capacity of Cushing's tank farms is nearly 100 million barrels. Say, posits Bridgwater, Ripple's scenario of losing half the tanks came to fruition.

"That's fifty million barrels," he says. "We'd be looking at our own on-land *Exxon Valdez*."

Worse, actually. The *Valdez* was carrying just over a million barrels of oil. It lost just a quarter of that. And light crude, like that stored in Cushing, poses particular challenges to an environment, often killing animals or plants on contact and emitting dangerous fumes that can kill both human and animal residents.

"This would not be a simple cleanup," says Bridgwater. "You'd have an uninhabitable community for a long time."

—✳—

In the late 1960s, after the earthquake events began at the Rocky Mountain Arsenal, the USGS staged a controlled experiment at an oil facility managed by Chevron in Rangely, Colorado. They wanted to see to what extent induced quakes could be controlled. What they found was that the tremors were in fact controllable by limiting or ceasing water injection. That's largely what happened in Youngstown, Ohio, too: Once the well ceased operation for good, so did the earthquakes. But Denver wasn't so lucky. The quakes provoked by the Rocky Mountain Arsenal continued until 1981—fifteen years after pumping ceased. Experts aren't sure why that was the case, but they speculate that it might be because of the sheer volume of fluid that was injected. In the case of the Dolores River project, the Bureau of Reclamation found that the longer underground pumping occurred, the less able they were to correlate injection rates and seismic events.

That could be bad news for states like Oklahoma, which have made other people's wastewater a cottage industry. In fact, right now, more water injected in Oklahoma comes from out of state than from within. "Oklahoma is pretty much an intake state," Matt Skinner, spokesperson for the OCC, explained to the Associated Press. The same is true for Ohio, which regularly injects water trucked in from places like Pennsylvania and West Virginia. Right now, Ohio has nearly 2500 wells, and they've received more than 202 million barrels of wastewater, over half of which have come from Pennsylvania.

Most of Oklahoma's water comes from Texas, which is kind of ironic, given Texas has its own growing earthquake problem. Unlike states such as Ohio, Colorado, and Arkansas that have placed growing restrictions on how and where wastewater can be injected, Texas currently operates on a policy just a few steps away from anything goes. If anything, argue public health experts, the locations of the state's 54,000 injection and disposal wells often seem related to race and economic class, with a disproportionate number located near poor and minority communities—a clear example, they say, of environmental injustice. That's easy for me to believe. The wastewater injection site I visited outside of Stillwater was located directly across the street from two tiny ramshackle houses. Both had their blinds pulled tight and an assortment of discarded appliances on the lawn. Tucked among them were well-used toddler toys: a tiny plastic slide, an even tinier wading pool, a torn playpen.

It's a scene replicated again and again in states like Oklahoma, where residents are often so busy just trying to get by that they don't have the time and resources to fight the siting of these hazardous facilities. Factor in that they live in areas just starting to get reliable Internet service, and it's easy to understand why they may not even know that such wells have been proposed. Angela Spotts says she thinks energy companies bank on this state of affairs, along with the lack of influence wielded by flyover states.

"We're Oklahoma. No one cares enough to get involved," she says.

Regardless of where the wells are sited in their state, Texas oil and gas companies there continue to fight any correlation between recent earthquakes and their wastewater disposal. After a series of damning studies refuted those claims, the Railroad Commission of Texas, which oversees the wells, changed permitting rules such that the commission could control what goes on at a particular well. As of this writing, they've yet to do so.

They also continue to deny the science. In 2016, a group of researchers at Southern Methodist University published the most definitive study on Texas's earthquake problem to date. The article, which appeared in *Seismological Research Letters*, showed a direct correlation between wastewater injection and many of Texas's recent earthquakes. A spokesperson for the commission issued a statement calling the study "arbitrary" and "purely subjective in nature."

The USGS disagreed. So, too, did the Environmental Protection Agency. In the summer of 2016, they sent a letter to the Railroad Commission expressing concern over the increased seismic activity in North Texas and reminding the commission that the connection between wastewater injection and induced quakes is "irrefutable." The letter continued, "EPA is concerned with the level of seismic activity during 2015 in the Dallas/Ft. Worth area because of the potential to impact public health and the environment, including underground sources of drinking water."

The Army Corps of Engineers issued a similar report. They petitioned for a 3000-foot buffer zone around Corps-maintained dams, arguing that induced quakes there could damage the structures. In a letter to Texas officials, they warned that activity associated with one particular site could cause "a catastrophic dam failure" if it were allowed to continue.

Nevertheless, some energy companies continue to bank on the opportunities that a lack of specific correlation or scientific certainty affords. In 2016, a group of Ohio lawmakers attempted to pass a bill regulating wastewater injection in that state. It was killed on the floor. Opponents cited the uncertainty of future events as a key cause.

This is the kind of response that really rankles Todd Halihan. He says a lack of total certainty never used to be a sticking point when it came to making safe choices based on the best science. We're always going to have some inherent uncertainty when it comes to induced seismicity, he says: "That's how this problem works. We have variabilities concerning wells, concerning pressure, concerning emerging science about faults." That's nothing all that novel, he says. Instead, what is new is what we do with that uncertainty.

"We used to believe that a perspective of uncertainty would be a reason to slow down something. Now uncertainty is being used to avoid things. Previously, you would have said, 'There's potential for a problem, let's stop what we think might be a problem, or alter what we're doing.' But now we use the uncertainty to avoid collecting data or trying to make it better."

The EPA could take over control of any state's wells if it wanted to. But that wouldn't solve the larger problem. For over a century, we have been a culture quite literally driven by oil. Increasingly, natural gas is playing an integral part in our lives as well. It cooks our food and dries our clothes. It heats our homes, particularly in rural areas where other means often aren't available. We hear it is a cleaner alternative than oil. In Maine, where I live, city buses running on natural gas are painted bright blue and green and proudly tout that they "run on clean natural gas." L.L.Bean, our state's beloved outdoor outfitter, prides itself on running natural gas buses throughout Acadia National Park. And natural gas is marginally safer than petroleum, particularly in places like China and Brazil, where the overwhelming number of vehicles on the road are creating serious health problems for residents. (Both countries, incidentally, boast that over 1.5 million vehicles on their roads are powered by natural gas. They pale in comparison, however, to Iran and Pakistan, which both have over 3 million such vehicles.) We're even using natural gas to fly planes.

For all this innovation, we still lack the means—in terms of both money and science—to map out the faults that may be activated by obtaining this gas.

Other so-called sustainable energy sources are posing similar problems for the stability of our landscapes. Take geothermal energy, for instance. It's clean, it's efficient, and it's becoming an increasingly practical alternative, particularly in places like Southern California and Iceland. That's in large part because of the tectonic activity there. Many geothermal projects are built on tectonic boundaries where faults allow access to the tremendous heat produced by the mantle. Plants there inject water to open up fractures and heat the water, which is then used to generate heat and electricity. By their very nature, these plants must maintain a delicate balance, encouraging rock to slip but not rupture. There's a certain degree of trial and error involved.

The area around Southern California's Salton Sea is one of our largest geothermal facilities. At first glance, it's a desolate landscape marked by dusty roads and boiling mudpots, the latter of which are as vivid as their name suggests—boiling caldrons of sulfur and mud that can build as tall as a massive termite mound.

Abutting these features stand massive facilities comprising rusting storage tanks, miles of pipes, and a lot of barbed wire. Plumes of steam rise from them—which seems impossible, given the sweltering temperature of the surrounding air. The plants themselves are known as flash-steam facilities, which is to say that they extract hot water from the ground, convert it to steam to propel energy-producing turbines, and then return the water to the ground. Some of the water evaporates along the way, so the water table below-ground gradually decreases over time.

Scientists have found a direct correlation between the activity of the plants and seismic activity in the area, which raises interesting questions about the degree to which an increase in overall fluid volume is responsible for induced earthquakes. At the other end of the state, residents in Napa Valley attempted multiple times to block the expansion of a geothermal facility there, saying they didn't want earthquakes. The site, known as the Geysers, was actually the first geothermal plant in the country, beginning energy production as early as 1922. But that was more of a quaint curiosity than an enormous industry, which is what it had become by 1987. By then, demand had far outstripped the extractive potential of the field. And so power companies began injecting local wastewater 10,000 feet into the ground, where it would be rapidly heated by the geothermal vents there—just like at Yellowstone. By 2001, it had become the largest steam producer in the world. It also became the producer of tiny earthquakes—more than double what there were before the wastewater injection began. But they've remained tiny, and so production has continued—despite the uneasiness of some residents. It also prompted a page on the USGS website addressing the frequently asked question "Why are there so many earthquakes in the Geysers area in Northern California?" Well, you know the answer.

Geothermal energy in Basel, Switzerland, has taken a very different course. The plant there was intended to provide electricity for 10,000 homes but was shut down after a series of small earthquakes, the largest of which was an M 3.4 (significantly smaller than the largest yet recorded at the Geysers, which was an M 4.5). Some properties in the area experienced minor damage. But the plan, which had the plant injecting water as far down as three miles, was deemed too risky. It didn't help when Markus Haering, the project head, told *The Guardian* that he had "very little knowledge of seismicity" beforehand and was surprised by the intensity of the quake. He subsequently stood trial. A Swiss court acquitted him, stating that they found no evidence Haering behaved recklessly or knowingly damaged properties.

Alan Morris has spent his life studying seismicity, particularly the kind induced by activities like geothermal power generation. He says geothermal plants always involve a kind of dance, generating as much power as they can but not *so* much that they set off an earthquake. He sees the Geysers as a success story. He thinks Basel might have been, had it been executed differently. He also thinks the residents there have a much lower threshold for what they'll accept.

In 1356, the area witnessed a quake believed to have been anywhere between M 6.0 and 7.1. It was the largest to occur in Central Europe, destroyed all the churches and castles in the area, wreaked considerable damage in what is now France and Germany, and prompted a massive fire that killed dozens and destroyed whatever buildings weren't knocked down by the tremors.

"Communities have a very long collective memory when they experience something like that," says Morris.

It's also worth noting that, like in the Rocky Mountain Arsenal area, earthquakes around Basel continued even after the plant was closed down.

The question that we need to be asking as communities, then, is how much are we willing to bear? And who gets to make that choice? Certainly the people I met in Stillwater would have been delighted if a paltry 3.4 quake had shut down operations in their area. Then again, some of the workers I spoke to at the Cushing tank farms said they're willing to take the risk. They said the only other viable job opportunity there is working in the prison, and they felt like they were a lot more likely to be killed by an inmate than an earthquake.

These questions aren't going to get any easier as technologies to counter global warming develop. Consider the innovation known as carbon sequestration. A few years ago, a consortium of private companies, federal agencies, and state-associated institutes successfully injected 1 million metric tons of carbon dioxide into a sandstone formation just outside Decatur, Illinois. That's the equivalent of the total CO_2 produced annually by an average coal-fired power plant. The idea here is that putting it deep underground, rather than releasing it into the atmosphere, will slow or maybe even reverse the effects of climate change. It's an increasingly popular initiative being attempted around the world in countries as varied as Algeria and Australia, South Africa and South Korea. In the case of the Decatur project, geologists captured CO_2 produced by Archer Daniels Midland, which emits large quantities of ethanol nearby. They dehydrated and compressed the CO_2, then injected it into a 7000-foot well. It then seeped into the pores of the sandstone and was covered with a layer of shale to keep it buried underground.

The USGS installed a seismic network consisting of twelve stations there and found clusters of microseismicity that accompanied the injection, but much of it was of a magnitude too small to be felt or to disrupt the shale cap.

Some critics say that a test facility doesn't accurately represent the conditions of a large-scale commercial project. And it remains to be seen to what degree CO_2 will behave like water when forced underground. What is known is that the CO_2 increases fluid pressure and can trigger both induced seismicity and other ground deformations. Lab simulations have shown that the process is capable of rupturing faults within a few days of injection. The National Academy of Sciences has come out publicly to say that larger-scale carbon storage projects may induce larger seismic events. And they have the potential to be greater than those seen at oil and gas facilities, since the injections are not matched with a similar withdrawal of fluids. The commercial facility in Algeria has triggered over 9500 earthquakes, though, again, most of them were too small to be felt.

Are we okay with small tremors that might shake our houses but not damage them, that might boom loud enough to interrupt our sleep but also keep our homes warm? What risk of a major quake is too great? And for whom?

These were the questions I found myself asking as I left Stillwater. On my way west, I stopped in the tiny town of Guthrie. It's a charming place filled with Victorian homes, a tiny BBQ joint called Stables, and a robust farmers' market. It was a Saturday morning when I rolled through town, and a statewide band competition was under way at the Guthrie Scottish Rite Temple, a massive stone-columned structure that looks like it belongs on the National Mall in Washington. The earthquakes in Guthrie are getting bigger. And as I sat, watching kids stream into the building with their trombones and clarinets and trumpets, I kept thinking of something Todd Halihan had said to me over dinner. That if the big one happens there, officials aren't going to be talking about casualties in terms of the few who died in the quake, but rather the few who lived. A lot of the families here are barely getting by. Relocating somewhere else isn't an option. Would these kids make it?

COOL STUFF TO STORE IN MINES

▌know what you are thinking: We've had such great luck pumping
▌water and military waste belowground, why don't we also store
some of that awful nuclear waste we have lying around—especially
the kind that has a half-life of 20,000 years or more—down there?
That's the kind of thinking that would have landed you a government
job thirty or forty years ago. We first decided to go down this road in
1976, with the inception of the National Waste Terminal Storage
program (NWTS). A feasibility study of sorts, it brought with it the
added bonus of the first real attempt to study seismicity in our subterranean structures. The thinking here was that if we could determine
that places like Lucky Friday were immune to major earthquakes, we
could plug them full of our nuclear waste and forget about it.

Dealing with that waste, particularly the thousands of metric
tons accumulating at commercial power facilities each year, has always been a problem for the industry. By the time the NWTS program was launched, we had managed to rack up over 40,000 metric
tons of it. According to a Department of Energy report, we bandied
about a series of solutions, including burying it (1) under the ocean,

(2) in the ice sheets of Antarctica, or (3) "beneath an uninhabited island that lies in a remote area and lacks natural resources." We also considered relocating it to the moon and "injection of waste into the sun." Remarkably, none of these solutions proved viable. And so the National Waste Terminal Storage program was the first real attempt to tackle the problem. It was renamed the Civilian Radioactive Waste Management program in 1983. No one can tell me why. I can't help but feel like replacing "National" with "Civilian" was an attempt to shift responsibility for the program. But I can't substantiate that. An early federal report stated that "although it was clear that a large-scale field test would be ideal for demonstrating essential technologies and revealing unexpected effects of waste emplacement, it was also evident that there would be few opportunities for such testing early in the program." And so, instead, we settled upon a much smaller project called the Spent Fuel Test—Climax. Wes Patrick was involved with the project from the start.

Patrick didn't set out to become a spent fuel expert. His background is in mining engineering. He went to work at the Lawrence Livermore National Lab, just southwest of Berkeley, one of the nation's two nuclear weapons testing laboratories (the other being Los Alamos). Patrick's plan was to work on alternative energy sources. Instead, he found himself at the Nevada Test Site working on the Spent Fuel Test.

Patrick says the Test Site was a logical choice: It'd already seen a lot of seismic study from the nuclear testing going on there. And since those tests were continuing, they knew they'd see even more. Engineers constructed a facility 1400 feet underground, first by using a nuclear test shaft about 1000 feet down, and then deepening the borehole with a hammer drill—a massive five-foot-long bit powered by compressed air and capable of shattering even the hardest rock. As compressed air drove the hammer drill, equally compressed water forced the resulting rock chips to the surface, leaving an empty tunnel. Engineers then lined the area with steel pipe cemented in place and stocked the area with spent fuel shipped from the Turkey Point nuclear facility in southwestern Florida. Once it arrived in Nevada, they encapsulated it and stacked it in the underground facility to see what would happen. Of particular interest to them was what the heat from the spent fuel would do to the rocks, particularly quartz, which can easily degrade and even slough or crumble under high temperatures. (That, Patrick says, is why countries like Sweden and Finland tend to focus almost exclusively on storing waste that has cooled for at least twenty years.)

Patrick wouldn't tell me exactly how hot this fuel was—just that it was really hot. Though he did say that fuel cooled for twenty years gives off heat in the "light bulb" range. His main question was how to take care of that fuel as it cools from really hot to light bulb to room temperature. For him, the issue wasn't just whether the facility would remain geologically stable, but also whether they'd be able to safely retrieve the fuel canisters if need be. The instrumentation used to measure stress change in the facility failed shortly after the project began—the "vagaries of the Nevada Test Site's power grid, shaking from nuclear weapons tests, and high temperatures caused by the thermal load" were too much for them, Patrick wrote in a study published not long after the completion of the test. Still, they had enough data to conclude that the basic idea was a sound one.

The Department of Energy designated ten locations as potential sites for deep geological nuclear storage; in 1987 then president Reagan commissioned an in-depth study at three: the Nevada Test Site, along with the Permian Basin (an underground salt bed in Texas) and a basalt formation in Washington State. The basic thought here was that the earth had held mineral ore for millions of years. Why couldn't it also hold nuclear waste?

But ensuring the stability of these sites presented formidable challenges. Geologists in this field spend most of their time looking for evidence of faulting and seismic activity—it's a process identical to the one Ben Henderson at Lucky Friday and other mine geologists use on a daily basis: mapping veins and the broken rock of a fault as it snakes underneath the landscape. What the Department of Energy and the Nuclear Regulatory Commission wanted was the exact opposite: evidence that there were no such features. So how do you do that?

That's the question Patrick's firm, the Center for Nuclear Waste Regulatory Analyses, part of the Southwest Research Institute, undertook along with the USGS. The three finalist sites were chosen in large part because they were in the middle of nowhere. That's an attribute that tends to appeal to American voters and taxpayers—at least where nuclear waste is concerned. But middle-of-nowhere places tend not to be hotbeds for geological inquiry, and so next to nothing was known about them: no academic studies, no historical accounts of earthquakes 300 years ago. That meant the geologists had to pretty much start from scratch. They quickly determined that tried-and-true methods like digging boreholes to take rock samples were no good: The boreholes, just like mine shafts, redistributed pressures on the surrounding rock and could conceivably create weakness where there had been none. Plus, it's one thing to seal up an old silver mine; it's quite another matter to cap a hole in the ground where nuclear waste might be stored. Even if it didn't heighten the risk of seismic activity, a borehole—even a sealed one—could allow groundwater to circulate, taking with it radioactive material. The USGS had been experimenting with different seals, but it would need another 25,000 years to make sure those seals would survive the half-life of elements like plutonium.

And so, instead, they tried relying on fancy geophysical processes like measuring rock gravity and magnetism. They installed as many seismometers at the sites as they could access. They picked up seismic activity around the Permian Basin believed to be caused by oil drilling there. They found faults and evidence of volcanic activity around the basalt field in Washington, but they couldn't say for sure whether they were active. Along the way, they worried a lot about quality control in their studies. An optimistic author of a USGS paper on the subject concluded that such work offered a "unique challenge, as well as an opportunity for geophysicists to think carefully about their craft and to document it adequately. Apparently, the challenge can be met, but at considerable increase in the amount of attendant paperwork."

Meanwhile, elected officials and lobbyists at each site began playing a passionate game of "not it." Nevada lost handily to Washington and Texas. In the meantime, the federal government commissioned studies at mines like Lucky Friday to understand just how stable (or not) similar locations might be. But costs for the feasibility studies at the three proposed locations soon became prohibitively expensive (over a billion dollars per site, by some estimations). And so Yucca Mountain, located at the Nevada Test Site, soon became the sole focus of the Department of Energy. A 1200-foot volcanic mountain located less than a hundred miles from Las Vegas, the ridge already belonged to the US government. Geologists focused their attention on two sections of barren land located on the edge of a shallow and sprawling basin named (and I swear I am not making this up) Jackass Flats.

I stopped there after leaving Texas and Oklahoma. Or at least I tried to. Yucca Mountain is over a two-hour drive from Las Vegas through scrub desert, past abandoned cabins and boarded-up buildings that look very much like Wild West saloons, to the land currently known as the Nevada National Security Site. The field office there offers limited tours once a month (though not to Yucca Mountain proper). My visit didn't happen to coincide with that tour, and the polite woman on the phone informed me that private tours are only arranged for groups and organizations of twenty-five people. She also made a point of informing me that cell phones, computers, recorders, data storage, laptops, "optical instruments," and alcohol—she emphasized that last one—are prohibited on the tour. Apparently, the reputation of writers precedes us. When I snickered, she emphasized that random searches for these items would absolutely occur during our tour.

Without being able to ply them with alcohol or optical instruments, it seemed unlikely I could quickly make twenty-four friends in the greater Las Vegas area willing to join me on a group tour.

Instead I decided to try my chances at the gate. Unremarkably, it didn't work. And so I didn't get to drive through the area known as Jackass Flats to visit the mountain itself. But that's okay: I was close enough to see that the land is very, very dry and very much at the base of the Basin and Range.

These lands were the site of intense nuclear testing during the 1950s—the same nuclear testing that prompted the international seismic network. The Department of Energy argued that this location—already the site of intense nuclear activity and in an isolated, dry part of the country—was ideal for nuclear storage. Nevada residents disagreed.

Nevertheless, exploratory work began there in earnest in the 1990s. Contractors began digging a tunnel five miles long and twenty feet wide into the mountain and the ground beneath. They spent, by some counts, billions of dollars studying the rock as they went: Would it leak? Crumble? Quiver?

The answers to these questions depend entirely upon who you ask. Officials representing Nevada were quick to point out that the state is the third most seismically active in the nation. They've already mapped thirty faults in the area immediately around and under Yucca Mountain and are certain there are many more. In the years since 1976, seismologists there have registered more than 600 earthquakes in a fifty-mile radius of Yucca, not including those seismic events prompted by nuclear testing. That includes the M 5.6 quake that occurred in 1992 on a previously unknown fault just eight miles from Yucca. It was strong enough to damage the newly erected Department of Energy outpost there. "If you're trying to find a good, stable geologic foundation for a repository, it's not available at Yucca Mountain," insisted Bob Loux, then the executive director of Nevada's Commission on Nuclear Projects and a longtime contributor to the US Atomic Energy Commission.

Geologists at the University of Nevada released similar findings, stating that the area is defined by "notable earthquake activity" and "young volcanic structures." They also pointed out that the trains carrying the waste to the storage site would traverse the country, passing through some of our most active seismic zones—and densely populated counties.

The Department of Energy never disputed these facts, mostly because they insisted the facts had little bearing on the stability of the repository. Underground structures, they said, are great at withstanding seismic activity. "And, in actual tests at the Nevada Test Site mine tunnels have withstood ground motion from underground nuclear explosions that are greater than any ground motion anticipated at or near Yucca Mountain. Repository facilities at the surface also can be designed to safely withstand earthquake effects."

Fuel rods, of course, are not the same thing as human miners. And no miner is required to stay belowground for 10,000 years or more. Mindful of this, the Department of Energy began focusing their research on titanium canisters capable of containing the rods and other waste, regardless of what happens around it. Gary Taubes, author of *Nobel Dreams* and *Bad Science: The Short Life and Weird Times of Cold Fusion*, wryly noted that the canisters were safe enough to toss under Boston's Fenway Park. No word from the Red Sox on whether they're taking up the initiative.

And while the debate between geologists in the Department of Energy and those in Nevada has yet to be settled, the controversy was enough for the US government to halt the Yucca Mountain project in 2011.

Nevertheless, the World Nuclear Association, an international organization that promotes nuclear power, remains strongly in favor of underground storage. They point out that deep geological storage facilities are in use in countries like France, Norway, and Sweden. They also note that the United States is already storing some of its nuclear waste underground. They are correct on both fronts.

WIPP is the somewhat unfortunate acronym for the Waste Isolation Pilot Plant project, another US Department of Energy nuclear waste initiative authorized by Congress. It's also the acronym for

Women Impacting Public Policy, a very different initiative and one that, as best as I can tell, has not been authorized by Congress.

The nuclear WIPP was authorized in 1979 and opened for business in 1998. Like Yucca Mountain, WIPP is also intended to store nuclear waste underground. Engineers designed it very much like an extraction mine akin to Lucky Friday, which is to say with a series of stacked ramps emanating from a shaft. Instead of a dormant volcano, though, WIPP uses an enormous bed of salt located below the surface. There, the Department of Energy trucks items like contaminated gloves and safety suits along with tools, debris, dirt, and anything else that got tinged with radioactivity during governmental operations. The official term for these materials is "transuranic waste." We've been accumulating it for over sixty years, and apparently it really stacks up. A salt bed located 2100 feet belowground just outside Carlsbad, New Mexico, seemed as good a place as any to put it. And so that area went through a geological survey much like the one at Yucca Mountain. Unlike Yucca Mountain, it got up and running and managed to accumulate radioactive materials for over a decade, until a series of accidents and mishaps forced its closure in 2014. The most significant of them involved a drum filled with kitty litter that had been used to absorb a radioactive spill at Los Alamos National Laboratory. Historically, lab workers there had always used synthetic kitty litter for this purpose, but they recently had switched to an organic version. Unbeknownst to them—and to officials at WIPP (and to most humans, for that matter)—that brand of kitty litter chemically reacts with plutonium and americium, two radioactive elements. The end result was an explosion of the drum, which spewed radioactive goo over one-third of the total area of the mine. Over twenty workers on the surface of the mine were exposed to radiation. According to the *LA Times*, the cost of the cleanup may exceed $2 billion, making it on par with efforts to repair damage caused by the Three Mile Island meltdown.

The area around WIPP does get its fair share of seismic activity—on average, about one earthquake a year measuring around M 3.0. They've had a few as high as 4.0 in the last decade. The USGS estimates there's about a 10 percent chance the area will see an M 5.0 earthquake in the next 500 years. Even if they're wrong, the Department of Energy is banking on the fact that salt is lousy at storing seismic strain; it's so weak, it tends to just crumble. But that can also make it susceptible to collapse. And the walls, floors, and ceilings of WIPP's underground storage areas are collapsing in on themselves at a rate of about 5 inches a year. While it was open and active, crews down below would install reinforcements and bolts much like the miners at Lucky Friday do: drilling holes in the ceiling and floor, then inserting long steel rods to keep the area stable. After the accidents, no one was allowed down below, in part because salt cannot be decontaminated. So for two years, no one has been allowed to go below to bolt up the storage caverns.

In the meantime, the Mine Safety and Health Administration has been assessing the facility, particularly with regard to mine stability. They found fifty-two violations during their assessment; the Department of Energy is working to address them. In their recovery plan, the Department of Energy noted that this kind of structural bolting is a high priority, even in the contaminated areas. They call it "catch-up" bolting. That's a problem, since the facility doesn't have the kind of protective equipment that would allow someone to work in such a contaminated space. They're currently "evaluating new personal protective equipment options in the marketplace."

The total cost of "recovering" WIPP is estimated at $242 million by one report. That same report indicates that mine maintenance is one of the major drivers of the total. It also notes that the recovery plan is based on "key assumptions" including: "1. Existing drifts and panels will remain stable and not suffer roof falls or cave-ins."

If history is any indication, things look good for WIPP. We've been detonating nuclear weapons in salt deposits for years, in part because they do a great job of concealing the blast. During the Cold War, the Soviets did, too. We know this because we spied on them every chance we got.

THREE

THE VOICE OF RAGE AND RUIN

BLOCKS AND BLOCKS

We may be a young nation, but we love antiquities. Those columns on the Guthrie temple can be seen on courthouses and banks and federalist homes across the country. Consider the US Capitol building: Its designers were so enthralled with classical architecture that they included not just one but all three types of Greek columns. The White House looks like it's straight out of ancient Rome. Towering over both of them is the Washington Monument. Few structures in this nation are more iconic, more directly representative of the seat of political power, than this 555-foot tribute to our first president. "The monument," boasts the National Park Service (which oversees the structure), "like the man, stands in no one's shadow."

Of course, it took a couple of tries to get the tone of such a tribute just right. The first attempt wasn't a building at all, but rather a large statue of Washington shirtless, his lower half loosely wrapped in what appears to be an oversized beach towel, enthroned and clutching a short sword. Congress kept it in the rotunda for two years, then decided they'd had enough of his strapping chest and so relocated it to the lawn outside the Capitol. It stayed there until 1908, when it was further demoted to the basement of the Smithsonian. It now resides in the National Museum of American History.

While all of this shifting was afoot, a new plan emerged to build the monument proper. This one, spearheaded by the Washington National Monument Society, eventually settled upon the current design, an homage to the great Egyptian obelisks and pyramids. Construction began in 1848 with a groundbreaking ceremony. Thirty-seven years later, at the monument's dedication, Representative John D. Long of Massachusetts read a speech written by Robert Winthrop, former Speaker of the House, and an attendee at the groundbreaking. In it, Winthrop prophesized the obelisk's future:

> The storms of winter must blow and beat upon it. The action of the elements must soil and discolor it. The lightnings of Heaven may scar and blacken it. An earthquake may shake its foundations. Some mighty tornado, or resistless cyclone, may rend its massive blocks asunder and hurl huge fragments to the ground. But the character which it commemorates and illustrates is secure. It will remain unchanged and unchangeable in all its consummate purity and splendor, and will more and more command the homage of succeeding ages in all regions of the Earth.

Since that speech, the monument has been struck by lightning dozens—if not hundreds—of times. So often has it been struck, in fact, that the copper and aluminum points on its cap have been melted down to sorry-looking stubs. It's seen its fair share of winters, too. But no storm has done even a fraction of the damage wrought by the 2011 Virginia earthquake. Centered nearly a hundred miles southwest of Washington, the M 5.8 quake sent the capital rolling. It did minor damage to the Smithsonian and knocked down three of the four spires on the National Cathedral (though the building itself remained largely undamaged). The quake was not nearly as kind to the Washington Monument.

Five hundred feet up on the monument's observation deck, the shaking began gradually: just a few wobbles—enough to make the on-duty rangers look around. But it quickly intensified, making it difficult for visitors to keep their footing as they tried to evacuate. Fifteen seconds later, chunks of the ceiling began to rain down upon the visitors as parents held little kids and crowded the entrance to the stairs. Luckily, no one was seriously hurt.

In the days that followed, surveyors rappelled down the faces of the obelisk, assessing damage. They inched up stairways and poked into corners. Much of the marble that makes up the pointy top part of the monument (known as the pyramidion) was cracked. Subsequent rainstorms left pools of water in the observation area and below. Chunks of mortar were missing from the interior masonry, and once-tight joints were so askew that surveyors could see daylight through them. Steel rods in the elevator shafts appeared bent, and exterior stone appeared broken or crushed in places.

Over the course of three years, construction teams added nearly three miles of sealant in between undamaged stones. They replaced damaged ones with hunks of old foundation and stairs salvaged from Baltimore row houses (the color was an almost exact match). They added stainless steel anchors and bolts. And then, with much fanfare, the monument reopened in May of 2014. Less than two years later, it closed again—this time "indefinitely" (though news in 2017 of a major new donor may change that). Persistent elevator problems made visiting its observation deck too iffy, said the National Park Service. They couldn't say for sure, but they suspected that moisture damage from the quake might have caused the mechanical failures.

Why did this iconic structure suffer so mightily when so many other Washington, DC, buildings escaped the earthquake with little or no damage?

To find out, I flew not to the nation's capital but, instead, some 2500 miles to West Hollywood, where they know a thing or two about earthquakes.

I arrive on one of those perfect LA days, when the weather brings elderly women to the outdoor market in search of lilacs and film executives in aviator shades actually smile at one another while wrangling over the last parking space on Rodeo Drive. A mile or so away, Daniel Zepeda and I are walking one of West Hollywood's main drags. To our right stand a parking garage, a bistro, and a retro movie theater. Across the street is an upscale grocery store. In ninety seconds or less, Zepeda can give me the life history of each building.

"See those medallions?" he asks, pointing to metal circles the size of pie plates in the side of one building. "That building has been reinforced." We stop in front of another that looks like it's made of brick. Zepeda's pretty sure it's just a facade. So we creep without looking like we're creeping into the alley behind the building. There, a dangerous-looking tangle of wires and pipes runs up a wall made of cinder blocks. "That's what I figured," he says with a grin.

Zepeda is a structural engineer and, by many accounts, one of the best in the business. As he was growing up in Mexico, his mother always told him that successful men were either doctors or lawyers. He wasn't sure he really wanted to be either. Then, in 1985, an M 8.1 earthquake rocked Mexico City. The shaking lasted nearly three minutes. It crumbled hotels, schools, and factories. By the time it was finished, 3000 buildings in the city were ruined. Another 100,000 were in serious need of repair. Over 10,000 people were dead; many times that were injured or homeless. That's when Zepeda began to think that the world might need another building expert more than it needed another doctor or lawyer.

Today, Zepeda works for Degenkolb, an engineering firm specializing in earthquake-hazard assessment and mitigation. It's a thoroughly twenty-first-century office, accessible only by key card (there are no buttons on the high-rise elevator). Almost all the employees are under the age of forty, and they hail from just about every continent on the planet. When they get stuck on a design problem, they ride their hoverboards around the cubicles or play a quick game of Ping-Pong while eating an endless supply of red licorice, available throughout the place in big bulk containers.

In late 2015, the firm was hired by West Hollywood to conduct an exhaustive inventory of the city's approximately 6000 buildings. Their mission? To catalog the buildings based on structure and, more specifically, how that structure might fare in a major quake. That's no easy task. The 2008 ShakeOut report predicted that, of the steel high-rise buildings built prior to the Northridge quake, about five in the greater LA area would collapse altogether. Another ten or so would be damaged so badly they would have to be demolished. A couple of dozen would have damage significant enough that the buildings would have to be evacuated and closed for repairs. A typical LA high-rise has about 1000 people in it on any given workday. The authors of the report couldn't say for sure which buildings would stand and which would fall—they said the kind of structural analysis required for that sort of thing was well beyond the scope of their work.

That's where people like Zepeda come into play. Overwhelmingly, most deaths, injuries, and damage in any given earthquake are caused because of the failure of man-made structures. Whether it's a high-rise, a classroom, or even the bookcase next to our bed, that's the most likely site of our undoing.

Consider the two massive quakes of 2010. One, an M 7.0, took place in Haiti; the other, an M 8.8, in Chile. The latter was 500 times more powerful, and yet its death toll was just the tiniest fraction of Haiti's (less than 1 percent by some counts). Why? One major reason is Haiti's lack of building codes. Most of the structures there were constructed without any real attention to how they would perform in an earthquake. It's not that Haitians have a disregard for quakes; it's often that they don't have the cash to make their homes and buildings safe. Seismic considerations are expensive, and they become even more so in tightly packed areas like the capital, Port-au-Prince, which has a population density of over 70,000 people per square mile (compared to Chile's 23.89). At the time of the quake, Haiti employed exactly one earthquake engineer.

West Hollywood has a lot more capital, but that doesn't necessarily solve the problem. Most of the buildings there are residential. The city wanted to know how those buildings might fare in a quake, but they didn't want a lot of disruption for the occupants. So Zepeda and his team are mostly confined to sidewalks, parking lots, and alleyways (although, once or twice, we do duck into a bar or café to get a better look at the structures in place there). It's a process code officers and engineers call "sidewalk surveys" or "rapid visual screening." FEMA has even devised a two-volume handbook dedicated to the subject. There is no rapid way to finish reading it, but if you muddle through to the end, you'll be scoring vertical irregularities and lateral redundancies with the best of them. The higher the score, the more likely it is a building will not collapse during a quake.

How particular buildings fare may surprise you. During the 2010 Chile quake, older indigenous buildings performed quite well, while more modern structures, like the Santiago airport terminal, sustained major damage and some high-rises failed catastrophically. During a recent quake in Italy, some medieval churches were left standing near a college dormitory built in 1965. The dorm collapsed, killing eleven students sleeping there.

Still, says Zepeda, you can draw some basic assumptions. Think back to gravitational forces and dams. Those forces behave the same way on buildings. Physicists and structural engineers talk about them in terms of peak ground acceleration, which is to say the amplitude of acceleration (the rate of change of speed) of ground motion. This, too, is recorded in terms of gs.

The 1994 Northridge quake in California recorded 1.92 gs as far away as five miles from the epicenter. It takes a lot less to wreck a building—as little as 0.2 g in some cases. That's in part because, like dams, a building is often wrenched in multiple directions either in quick succession or even simultaneously.

"Twist," says Zepeda. "That's what really kills buildings."

Remember, too, that the g-force gets multiplied by the building's mass. That's one reason why houses made out of wood and other light materials tend to fare better than those made out of stone.

Inch for inch, wooden houses weigh a lot less than stone ones. Put both of them in the exact same earthquake, and the wooden one will do much better. The heavier the building, the more force in its movement, and that makes it all the more likely that some key component will break, buckle, or bolt. This assumes, of course, that all other factors between the wooden and stone house are equal—and they very rarely are. Soil type makes a huge difference in terms of how a building experiences shaking. One reason why the 1985 Mexico City quake did so much damage was because the city, like much of Salt Lake City, is built on old lake beds. Soil like that can liquefy and become unstable. The height of a building matters, as do any design features that create asymmetries or other weaknesses in a structure.

Zepeda hasn't studied the Washington Monument specifically, but he says he suspects that its height and slender dimensions were big reasons why it sustained so much more damage than surrounding buildings—that the top of the structure had a whip effect not unlike what happens at the end of a water hose when you turn the water on full blast. Certainly, he says, it and most other East Coast buildings were built without seismic consideration in mind.

Across the country, we've had very few opportunities to observe how big steel buildings, like most skyscrapers, perform in earthquakes.

But what we know, says Zepeda, leads to a "spectrum of damage possibilities." The more ductile they are, the better they tend to withstand swaying and shaking. During the 1995 Kobe earthquake, steel buildings fared pretty well. But, Zepeda warns, that's in part because Japan has a much stricter building code than we do.

Evaluating how each of these factors comes into play in a city requires a process that is one part French flaneur, one part forensic detective. You have to look closely while also looking subtle. The FEMA guidebook recommends carrying binoculars, a camera, and a clipboard at all times. You might be able to get away with the first two in some parts of Hollywood, but the clipboard will give you away every time. Zepeda relies on his cell phone quite a lot. No one ever thinks twice about a young well-dressed guy looking at one of those.

On our tour of West Hollywood, we first stop in front of a classic storefront—the kind with plate glass windows and attractive wooden detail work around the doorway. To my eye, it is a perfectly fine structure—tidy and solid. But Zepeda is concerned about the irregularly sized bricks used in its foundation: They're old, he says. Bricklayers have been relying upon standard-size mass-produced bricks for decades. The lopsided ones used here tell him the building was completed long before engineers understood how structures were impacted by ground shaking. And that means it also might not include crucial rebar or other supports. People in the business of earthquake safety have a term for this kind of construction: "unreinforced masonry," or URM. It is the bane of their existence.

URM buildings consistently show more damage than other types of modern construction in large part because their walls tend not to be fastened to their floors and roofs. As the building sways in a quake, the wall ultimately pulls away from the floor, releasing the pressure that held it in place. Much like we saw in mines, this can cause a floor or roof to collapse. And with it can come the entire building.

Even if an entire URM building doesn't collapse, it can send an avalanche of bricks and stone that often prove deadly. The 1983 earthquake that rattled the workers in the Salt Lake City airport tower resulted in two deaths: Both were kids crushed by falling brick and stone while walking to school. In 1886, Charleston, South Carolina, was built almost entirely out of brick. Over 80 percent of the buildings there were damaged when that quake struck. One of the main reasons why you won't find many brick buildings in Santa Barbara today is that most of them were wiped out in the 1925 quake there. Not long after, officials passed new building codes mandating the iconic Spanish style architecture for which the city is now famous. At last count, New York City had about 30,000 URM buildings. If an M 7.0 quake were to rupture, about 6500 of those buildings would either collapse during the quake or be so badly damaged that they would have to be demolished. Only about a fourth of the total buildings would survive unscathed. FEMA estimates about 6000 people would die in this scenario, with another 13,000 people injured. Over three-quarters of a million people would be homeless. And the total bill for repairing the damage to buildings alone would be about $11.5 billion.

That's just one reason why the current International Building Code prevents new URM construction in huge swaths of the United States: If you live in either the Pacific or Mountain Time Zone, don't even think about it. Ditto for a wide horizontal band stretching from Missouri to Virginia. New York? You're restricted as well. So, too, is the northwest corner of Maine, which, in truth, is probably only bad news for a handful of enterprising bears and moose. Other places, like Chicago, are allowed URM by the building codes but have nevertheless passed their own ordinances requiring inspection of URM buildings and other potentially hazardous construction.

Back in West Hollywood, Zepeda points to a row of small round medallions evenly spaced about ten feet up the storefront. They're painted the same color as the rest of the building and so easy to miss if you're not looking. To Zepeda, these metal plates, each no bigger than a coffee cup saucer, may well be the difference between life and death for the building's occupants. They're signs of a retrofit— that the once precarious facade is now bolstered by metal reinforcing that may be enough to allow occupants to escape unharmed.

The building next to it houses a dry cleaner. Zepeda asks me what I think it's made of. I guess "pebbled stone," not even knowing if that's actually a thing. He's polite.

"Maybe," he concedes. "But look up there." He points to a parapet rising from the roof. It's thin—too thin for stone.

"I'm guessing wood," says Zepeda.

Next, we duck into the alley behind a diner with an extensive wine list. The rear of the building is a mess of pipes and tangled cables for everything from electricity to Wi-Fi. Zepeda grimaces. They're bolted to aging concrete blocks—an untidy attempt to update an old structure. Across the street is an upscale supermarket. It's concrete, too—Zepeda is sure of it. "Look at the columns," he says. "No doubt."

Concrete holds a particular fascination for Zepeda right now. He's just returned from Pingtung City and Tainan in southern Taiwan. In February 2016, a magnitude 6.4 earthquake struck the city and surrounding region. The death toll was at least 116 people, most of whom died when the Wei-Guan Golden Dragon high-rise apartment tower folded into a pile of rubble. Ten other buildings collapsed, including a bank and a marketplace. The four bottom floors of the Wang Lin Hotel pancaked belowground; what remained leaned at about the same angle as the Tower of Pisa. Another 247 buildings were damaged beyond repair. Approximately 400 schools would open only after major renovations. Each one of them was constructed out of nonductile concrete.

Ductility is what allows a material to bend and stretch. Gold and copper, for instance, have high ductility, which is why metal-lurgists can roll them into thin wires or bend them into everything from circuits to hoop earrings. Concrete, on the other hand, has lousy ductility.

"When shaken," says Zepeda, "it tends to explode." He pauses. "Some engineers don't like the word 'explode.' Let's say 'breaks.' Like really, really breaks."

That's what happened all over southern Taiwan in 2016. Zepeda and his team were some of the first foreigners on the scene. They wanted to see what they could learn about concrete buildings and how to make them safer. They knew residents in the city were still pretty shaken up, so they brought suitcases of Ghirardelli choco-lates as gifts. I ask him if this was symbolic or just ironic: The San Francisco–based confectioner was one of the few businesses to sur-vive the 1906 quake; the signature dish at their store is still the earthquake sundae—with eight scoops of ice cream and eight dif-ferent toppings, it's sure to be "a family favorite," their menu promises.

Zepeda shrugs with a smile. He's more of a licorice kind of guy. And he's much more interested in talking about what they found on their trip. The Taiwan quake was a valuable one, he says, particu-larly in terms of intensity. In 1999, a much bigger quake—at least an M 7.6, though some scientists say it may have been even higher—wreaked all kinds of havoc on the island.

"That's what you'd expect," says Zepeda. "Even a strong build-ing is going to have a hard time in a quake that strong."

A much smaller quake shook the same region a few years later and left most of the buildings—even the ones with serious struc-tural weaknesses—undamaged. What happened in 2016 is what Zepeda calls the Goldilocks effect—just the right size to show which buildings are most vulnerable.

The seventeen-story Wei-Guan Golden Dragon apartment complex was one such building. In the days following the quake, accusations of shoddy construction emerged: namely, that its developer and contractor, Lin Minghui of Wei-Guan Construction Company, knowingly used subpar concrete and far fewer support beams than were required. In April 2016, he and four others were charged with professional negligence, a crime that comes with a punishment of no more than five years in prison.

But that was an extreme case. Other buildings that collapsed were constructed in good faith or based on the building codes enforced at the time. Following the 2016 quake, the Taiwanese government conducted an extensive study of soils on the island. They found that much of it was prone to liquefaction and that, as a result, many of the concrete buildings above it were in danger of collapse—about 4000 buildings in all, and many located in the capital, Taipei. The government earmarked $735 million for their retrofits.

Zepeda says more study will need to be done to determine precisely what brought down so many buildings in Taiwan. But he says he did learn some valuable lessons about what kind of reinforcement concrete needs (*horizontal* as well as vertical, he stresses). He says his team also saw some significant trends in what is called the torsional property of a building, which is to say its ability to twist without breaking. Most infuriating were what he calls the "mysteries of nature"—buildings that had obvious weaknesses but did just fine; a completely collapsed building right next a perfectly fine one.

"Sometimes all you can do is scratch your head," says Zepeda.

That was very much his reaction to a series of earthquakes that all but leveled Christchurch, New Zealand. The biggest occurred on September 4, 2010—a magnitude 7.1 event with an epicenter about twenty-five miles outside the city. But it was a subsequent aftershock—just 6.3 and a full six months after the main rupture—that flattened much of the city. By the time shaking stopped that February day, 185 people were dead, over half of whom were working in a concrete office building that collapsed and then caught fire.

An additional 6000 people were injured. Why? Seismologists think part of the answer lies in the fact that the epicenter of that quake was almost directly beneath the city. That particular rupture also resulted in extremely high ground motion (more than 2 gs). It was also a shallow quake, which means more of the energy could get to the surface. As a result, over 1000 buildings in the city's central business district—over 75 percent of all buildings there—were found to be so badly damaged that they would have to be demolished. So, too, would over 10,000 homes in the area. An additional 11,500 required significant repairs. In total, a nearly five-mile-square area of Christchurch was designated as a "red zone," which is to say that it was deemed uninhabitable. And, in the end, engineers found that 97 percent of the URM buildings there that hadn't been retrofitted were included in that zone. Bricks torn loose from those buildings were responsible for forty-two deaths. Early estimates for repairs came in at over US $32 billion. A year after the quake, Zepeda made his first visit. He says the city still looked "like a war zone." It took New Zealand disaster responders a week to install portable toilets for residents there, and they brought far too few to go around. When Zepeda visited, even most of the livable homes were without plumbing.

It was a powerful reminder, he says, about how it's often our modern first world buildings that fare the worst in earthquakes. Indigenous homes built of timber and bamboo have the kind of flex needed to weather seismic waves. A 2001 earthquake in northwestern India killed 20,000 and left another 600,000 homeless. Most of these casualties occurred in modern masonry buildings. Villages dominated by traditional building methods of grass and clay saw much higher survival rates—both for the structures and for the people occupying them. That's been true in places like Chile and Turkey, too.

Five years after the quake in Christchurch, many residents were still living in their cars or sleeping in church parking lots. Much of the downtown still hid behind tall demolition fences. Peter Townsend, a leader of the Christchurch chamber of commerce, told *The Wall Street Journal* that plans for the city to be up and running on the quake's fifth anniversary were "hopelessly optimistic." City officials say that one of the primary issues has been insurance. Although something like 90 percent of the residents there were covered, major delays with companies have meant that nearly half the residents have yet to receive a payment: The total settlement, which the Christchurch chamber of commerce chief executive likens to that of Hurricane Katrina, has tapped the reserves of even large insurance companies.

That's precisely the scenario West Hollywood wants to avoid.

Steve Bailey, the city's building and safety manager, says it's about creating a community where people can come back quickly. "We know that, in a big seismic event, West Hollywood isn't going to be the only place hit. We don't want to be competing with other cities for resources and support."

California has led the nation in seismic ordinances, particularly where building is concerned. In the 1970s, the state passed legislation banning new home construction in mapped fault zones unless contractors and potential homeowners could prove that the faults in question didn't prove a hazard. That era also ushered in a commitment to map the state's fault zones in detail. But money and personnel quickly ran short. Critics say the state has been slow to produce maps of the known faults. Just recently, the California Geological Survey issued the first updated maps in over twenty years. There were a lot of surprises for the nearly 1500 owners who discovered their property is vivisected by one or more faults. An additional 10,000 properties lie within the immediate hazard zones created by these faults. Ministers, preschool teachers, hotel managers, and nursing home workers all told *LA Times* reporters they had no idea their places were at risk.

Meanwhile, that same *LA Times* story reports that an estimated

2000 miles of known faults in the state have not yet been mapped. Construction on them has continued apace, even though they pose a significant threat to the occupants there. And these aren't just little beachside bungalows or ranches. They're high-rise apartment complexes and commercial buildings. And what we thought we knew about even the known faults is changing as well. Ever since the 1906 San Francisco earthquake, seismologists have speculated that the San Andreas Fault pretty much only ruptures in discrete sections. In 2014, the USGS, along with the California Geological Survey, updated their hazard assessments to include a brand-new doomsday scenario: the possibility that the fault could rupture along its entire length. Likely? No. But undeniably devastating. Were such an extensive quake to occur, over 3.5 million homes would likely sustain significant damage. The cost of repairing them alone would be an estimated $289 billion.

New Zealand and the United States have remarkably similar building codes. The earthquake that rocked Christchurch wasn't caused by a major fault zone. In fact, it occurred on a previously unknown fault. So, too, did the Northridge earthquake. And the Landers quake. The fact is that the overwhelming majority—nearly *every* earthquake that's occurred in this country in recent memory— have occurred on faults we didn't know existed. The threat posed by these unknown faults is particularly felt in places like the Northeast, where we know the least about the geological risk.

Zepeda has a pretty good idea which of the West Hollywood buildings will make it in a 6.4 quake, which ones won't, and which ones may but will have really expensive damage to repair. Part of the agreement I made with West Hollywood in exchange for detailing this survey is that I wouldn't report how Zepeda thinks any given building will fare. But suffice to say, there are plenty of buildings in all three categories.

He says that if you want to make a structural engineer happy, we'd all choose to build squat, perfectly rectangular bunkers—preferably without windows and doors. "Or maybe just one small door," he concedes with a laugh. "But nothing else."

Remarkably, that's not a style that has yet taken off in design circles. "No one is going to buy my bunker," he concedes. "People want views and stairways and parking garages. Every one is a potential design weakness." Pillars, garage doors, and windows all add structural weaknesses to a building. So, too, do any asymmetries. If you want to get Caltech's Tom Heaton swearing like a sailor, ask him about the building that houses his office. It used to look a lot like Zepeda's bunker. Just recently, the university installed a glass vestibule and staircase right outside the office of one of the world's leading experts on seismicity and structural integrity. No one thought to ask his opinion on the subject. That's too bad: He has a million reasons why it's a truly terrible idea.

The question every building owner has to answer is just how many weaknesses they are willing to tolerate. Do they want their building to remain fully functional after a quake? Are they content with just knowing that people can get out safely? Do they have the cash to help make that possible?

Answering these questions doesn't take into account the secondary loses that occur as a result of an earthquake. The greatest of them all is fire, which often does more damage than the quake itself.

Lucy Jones was one of the lead authors in a ShakeOut study to determine what might happen in Los Angeles in the event of a major seismic event. She says the 1906 San Francisco quake is not a bad model for what could happen to places like Los Angeles today. That quake did plenty of structural damage, but it was the subsequent fire that really leveled the more than 28,000 buildings destroyed there. Over 3000 people perished, either in the shaking or in the subsequent fire. Another 250,000 were left homeless. "Surrender," wrote San Francisco's beloved Jack London, "was complete."

To understand why, you have to also understand how water is stored and moved in LA. We've already seen how some of the dams there will fare. A much bigger problem, says Jones, is going to be the aqueduct that conveys the water behind those dams to the people of Los Angeles. In the ShakeOut scenario, over a mile of the aqueduct tunnels experience collapses and failures that dramatically reduce the flow of drinking water to the city. As a result, water would back up in the remaining tunnels, making work there an exceedingly dangerous proposition and one that could only be attempted by a select few divers. Repairs could take years. Until they are completed, potable water would have to be trucked into the city. With nowhere to go, water would continue to rise in the reservoirs, eventually flooding surrounding areas.

Meanwhile, predicts the report, damage to sewers and wastewater treatment plants could result in 10 million gallons of raw sewage pouring into the streets each hour at anywhere from fifty to a hundred locations: That's anywhere from 750 to 1500 Olympic-size swimming pools every hour. "Ultimately," concludes the report, "some of the damaged pipelines, equipment, and electronics take up to five years to repair."

But that's just the start, says Jones. An earthquake like the one envisioned in the ShakeOut scenario could also cause 1600 fires—even more if it's a particularly dry season.

The 1906 San Francisco quake broke almost all the water mains in the city. Even the few buildings with sprinklers burned down since there was nothing to power them. The very last cable sent before the telegraph lines went down read, "City in ruins. In flames. No water. Blowing up buildings to contain fire."

Much more recently, both the Loma Prieta earthquake in 1989 and a similar quake in Taiwan a decade later showed just how fragile our water system can be. Both saw significant damage to the water supply as pipes cracked under the force of the shaking. The problem, says Jones, is that these types of pipes have particularly thin walls for their diameters.

Without water, firefighters would be unable to extinguish blazes caused by the hypothetical LA ShakeOut quake. Anywhere between $50 and $100 billion in property—over 200 million square feet—would be reduced to ash. That's the equivalent of 133,000 homes. Those same fires could easily take the lives of a thousand people and leave another million homeless. The cost of insurance payouts would have sharp economic effects around the world.

As we turned toward this new century, FEMA challenged other communities to conduct similar analyses as part of a program the federal agency called Project Impact. The New York City Area Consortium for Earthquake Loss Mitigation, an association of universities, state agencies, and local and regional first responders, took up the challenge. Their multiyear, $250,000 study predicts an outcome no less grim for the Big Apple than for Los Angeles. Were an M 7.0 quake to strike New York City, resulting fires would create a demand for water fourteen times greater than the supply. The New York City Office of Emergency Management estimates as many as 900 fires could break out simultaneously. Without adequate water, trucks, or personnel to fight them, damage tolls could be as high as $15 billion.

It's hard to get your head around something like that.

When it comes to preparing for an eventuality like this one, Jones says, Americans suffer from a profound lack of imagination. The last remaining survivors of the 1906 San Francisco earthquake died years ago. When most Americans think about California quakes, they think about Loma Prieta—a mere 6.9—or the 1994 Northridge earthquake, which registered just a 6.4. The ShakeOut scenario is over thirty times stronger than either of those. It's difficult to comprehend the significance of that. The largest thermonuclear bomb ever made was the Soviet Union's Tsar Bomba. It had the explosive power of 3800 Little Boys—the bomb that was detonated over Hiroshima. In terms of total energy released, Tsar Bomba is only the equivalent of a magnitude 7.0 earthquake. We don't yet have an analogy for what an M 7.8 earthquake would look like: It's just too big.

—~~~—

If you want to see Daniel Zepeda really excited, ask him about historical earthquakes. He's a student of them the way Civil War buffs study Gettysburg.

"Every earthquake teaches us something important," he says. In a lot of ways, building codes are a testament to great quakes and the lessons learned from each.

The first major earthquake to hit urbanized America was the 1906 San Francisco quake. It destroyed more than 28,000 buildings and left over half the city homeless. A month after the 1906 quake in San Francisco, the Structural Engineers Association of California was formed. Their conclusion? Wooden structures fare much better than brick and stone ones—the dread unreinforced masonry. In emergency management and engineering parlance, "URM" is pretty much a four-letter word. That was true in 1906, too.

The mortar dust had not long settled when the Structural Engineers Association recommended that the city move away from brick and masonry construction during its rebuilding, but it met with fierce opposition from bricklayer unions. Their objections were paltry, however, compared to those lobbied by the Southern Pacific Railroad Company, whose very profits depended upon people and companies willing to come to San Francisco. In an attempt to reverse their plummeting stock losses, railway executives launched an elaborate PR campaign arguing that it was the subsequent fire, and not the earthquake itself, that damaged the city. Their passenger agent, James Hornsbaugh Jr., dedicated himself to the task, penning article after article in the railway's magazine assuring passengers and investors that seismicity wasn't a problem in the Bay Area. He also wrote to chambers of commerce around the country, beseeching them to cajole scholars and newspapers to focus on the fire rather than the quake. "Have them show how little an area the earthquake really effected," he pleaded. "How quickly the city and state recovered, that the burned city was an old city, and that only the most ancient part of the water-supply system failed."

The campaign worked: Little, if anything, changed in building construction. At least until 1925, when a magnitude 6.3 earthquake destroyed much of downtown Santa Barbara. City officials responded with the nation's first seismic codes, mandating that all commercial buildings be constructed to withstand future shaking. The best way they knew to do that was to dictate design elements: If you've ever wondered about the city's nearly pervasive Spanish architecture—those iconic white stucco walls and red tiled roofs—it wasn't nearly so much an aesthetic decision as it was a best guess as to what would survive a future quake. Other historic quakes prompted even bigger changes.

After the 1933 Long Beach earthquake, California passed legislation prohibiting the construction of unreinforced masonry buildings. Beginning in 1986, individual communities were required to identify and manage any risk posed by existing URM buildings. That same 1933 earthquake damaged nearly three-quarters of the schools in Long Beach. And so what also followed was the Field Act, which regulates all seismic considerations for the state's public schools: including their construction, renovation, and inspection.

Hospitals fared particularly badly in the 1971 San Fernando earthquake. In the newly opened Olive View hospital, three of the four stair towers pulled away from the building and crumbled, leaving only one way to evacuate patients. Several wings of the building sheared away from the main structure, which had been knocked over a foot out of alignment and was nearing total collapse. The first floor of the psychiatric unit pancaked, and surviving patients began wandering the grounds and beyond. Meanwhile, a canopy crashed into the hospital's fleet of ambulances, crushing them. All power was lost in the buildings, which meant evacuation—even from the highest levels—happened in near total darkness.

In the neighborhood of Sylmar, nurses and patients at the San Fernando Veterans Administration hospital watched in horror as four entire buildings disappeared in an instant. Forty-six people were killed. That destruction resulted in the hospital seismic safety legislation in 1973. It also prompted the USGS to install twenty-eight advanced earthquake-monitoring systems in VA hospitals across the country. They're in places you might expect, like California and Alaska, and in some you might not, like Oklahoma, Illinois, and South Carolina. The idea here is that they'll be able to analyze how the hospital performs. That'll help refine future code, but won't do much to make the current patients safe.

Two decades after the San Fernando quake, another earthquake in the region forced the evacuation of eight hospitals for reasons ranging from power loss and water main ruptures to more extensive structural damage. Phoenixlike, the Olive View Medical Center had managed to reopen as a new hospital—this time with new seismic safeguards in place. The building itself fared much better, but damage was still considerable (and unexpected): TVs bolted to walls flung themselves at patients, ceiling panels collapsed, water pipes exploded, and elevators went off-line. By the end of the day, administrators were forced to order the evacuation of all 377 patients.

This is part of what so captivates engineers like Zepeda. With the seismic design of most buildings, the goal is evacuating occupants safely, even if it means that no one will ever be able to set foot in that building again. The basic threshold is that there's reasonable assurance the building will neither collapse nor cause debris to rain down on occupants, passersby, or neighboring structures. With hospitals, the bar is much, much higher. Studies show that upward of 95 percent of people rescued in collapsed buildings are saved by fellow victims rather than emergency responders. Many hospital patients can't get themselves or others out of those buildings. And we need the units inside to keep functioning so that they can treat both the existing and the newly injured. That's hard to do.

In 2006, a moderate 6.7 earthquake shook the Big Island of Hawaii. At the Kona Community Hospital, the shaking from the tremors knocked over partitions separating beds and units. It ripped drop ceilings from their supports and sent individual tiles, along with fluorescent light bulbs, careening down on patients and staff. According to a subsequent FEMA report, it also dislodged "decades of accumulated dust . . . adding to the general disarray and threatening the health of the patients and other occupants" (not to mention probably totally grossing them out). Like Olive View Medical Center, the Kona hospital also lost power, though a backup generator was able to meet some of their more critical needs. That hospital also made the decision to evacuate its patients: Thirty-six of the most critical were transferred to other hospitals; twenty-seven were discharged and wished the best of luck.

California recently set a deadline of 2030 for all of its hospitals to be fully functional after even a major quake. It's a great goal, but many hospitals, especially in poor and rural areas, are having a hard time coming up with the funds to meet this requirement. Recent reports suggest as many as seventy of them still don't meet the state's basic seismic standards. A lot of them were built before the 1971 quake and don't include the design parameters that subsequent legislation required. That includes major facilities in the Bay Area. Skeptics say as many as half the hospitals there won't meet the 2030 deadline. In some cases, hospitals are choosing to throw in the towel rather than try to meet the new requirements. That's true in Berkeley, where the town's only hospital is planning to shut down as those new standards take effect—administrators there say it's just too expensive to try to meet the requirements.

In a recent self-authored report, the VA admitted that it failed to identify seismic risks for nearly 20 percent of its critical and essential treatment facilities located in high-risk seismic zones like California and New Madrid. They also said they have not yet mitigated nearly half of the structural deficiencies they did identify. They cited a lack of funding as the cause.

Still, study after study proves that keeping our hospitals functional will be one of the most important things we can do to save lives. According to the 2008 Los Angeles ShakeOut scenario, an estimated 50,000 people will have injuries requiring treatment. In New York, that number is about 13,000. Even on a good day, earthquake injuries tax the training of our best doctors and nurses. They crush legs, feet, and heads, rendering victims unable to get to hospitals of their own volition. That means we also need ways to transport them to hospitals. The LA report estimates at least 1000 people will need EMTs to get them to hospitals if they are to live. Burn victims also make up a shocking number of earthquake victims. In both the New York and LA scenarios, which have the quakes happening midday, those victims are likely to be mothers and kids as well as the elderly. Burns are notoriously hard to treat and can come with mortality rates as high as 51 percent. Survivors require special beds and intensive care units to tend to both their lungs and their damaged bodies. Currently there are ninety-two beds for burn victims in Southern California. The ShakeOut scenario has 564 people needing them. New York's most recent scenario study doesn't distinguish types of injuries, but a similar shortage would no doubt affect that city—perhaps to an even higher degree, given the increased difficulty people there may have exiting multistory buildings. And as hospital functionality decreases in the event of a major quake, the city could be looking at a total shortage of beds exceeding 1000.

Across the country, new faults are sparking additional structural concerns among residents. In 2016, ExxonMobil released new maps showing that faults run below both Dallas Love Field and Dallas/Fort Worth International Airport. They also run below the city's fabled medical district. If Heather Savage's theory of remote triggering is right, there's reason to believe those faults could become active as well.

Unlike California, Texas doesn't have much in the way of specialized seismic building codes. By most accounts, highly built-up cities like Dallas would incur significant losses in the event of even a moderate quake.

Fault lines run precariously close to six hospitals in Manhattan as well. That city also has about 13 billion square feet of buildings valued at about $1 trillion; 79 percent of those buildings are unreinforced masonry. A study conducted by the New York City Area Consortium for Earthquake Loss Mitigation estimates that in a moderate M 6.0 quake, 2600 buildings would be rendered uninhabitable: over 130 times the damage incurred on 9/11. Even a moderate quake, like the one that struck the city in 1884, would create at least as much debris as the 9/11 terrorist attacks. Were that M 6.0 quake to occur at 2:00 on a workday, an estimated 1170 people would perish. How likely is that quake? It has a 500-year return period. The 1884 quake has a 100-year one. You do the math.

So what's to be done?

In 2016, the city of Los Angeles adopted a seismic retrofit ordinance, which requires all buildings classified as "vulnerable" to be strengthened. That includes nonductile concrete buildings constructed before 1977 and even any wood-frame houses that have stability issues because of garages, columns, or other design elements. It's a similar ordinance to the one passed by San Francisco in 2013.

And researchers are busy developing new, more seismically resilient building materials using everything from laminates to coconut shells (the Obama White House even staged a competition for designers to construct earthquake-resistant multistory buildings). Daniel Zepeda is intrigued—maybe even optimistic about those kinds of technologies—but he doesn't want to wait for them. If he had his way, everyone would be taking pains to retrofit their structures now. But, like levees and dams, buildings are expensive to update. And for most of us, earthquake hazard is really pretty low.

Zepeda worries, though, that that kind of thinking overlooks the equally important idea of *risk*. Take New York City. The likelihood of an M 7.0 quake there is pretty infinitesimal—it has a return rate of 2500 years. So which quake do you build for? Answering that question has to take into account a lot more than likelihood. It also has to consider impact to people and structures. For instance, a 7.0 is going to do a lot more damage to Manhattan, New York, than it is to Manhattan, Kansas, simply because of the density of people and buildings.

New York City passed its first seismic building codes in 1995. Those codes only address new construction. A report issued by the New York City Area Consortium for Earthquake Loss Mitigation explains why: "The consensus opinion is that retrofitting thousands of New York buildings to meet seismic standards is impractical and economically unrealistic."

Amid much debate, Memphis made a similar decision in 2003, despite federal urging that they adopt building codes similar to those in California. A study offered by a team of university researchers and the Los Alamos National Laboratory concluded that the required retrofits would cost more than repairing the damage from a quake. The study was controversial, and many in the industry continued to urge stiffer codes. They tried to advance legislation to that effect in 2012, but it was defeated.

Other states and regions have their own riff on the codes. Oregon requires site-specific hazard assessments for what they call "essential facilities," including those that deal with hazardous waste, schools, and public works departments, along with hospitals, fire departments, and police stations.

In Massachusetts, communities renovating those essential buildings are required to include seismic updates. That's created problems for communities in and around Boston, where people say they need new facilities but don't have the money to make them earthquake safe. As one fire official told *The Boston Globe*, "We're just now getting our heads around the idea that we have hurricane risks here."

In Seattle, the Cascadia Subduction Zone has rightfully gotten a lot of attention, but it's actually the eponymously named Seattle Fault that poses the biggest risk to the city. M 6.0 quakes are pretty common there, and have happened, on average, about once every twenty years. The most recent one occurred in 2001 and did over $20 million in damage. Those quakes also manifest very differently than the kind of shaking that would be seen in a megathrust subduction event, and that poses additional challenges for both city building officials and design engineers.

Still, Seattle has made little progress toward adopting a more stringent seismic code. At best, their seismic retrofit could be called byzantine: Like in Massachusetts, the code requires a seismic retrofit only if a building owner takes out a permit for any renovation work. *The Seattle Times* recently obtained from the city a list of 1100 unreinforced masonry buildings that have not undergone any kind of seismic retrofit. That list includes schools and retirement centers, apartment complexes and doctors' offices. When the next earthquake happens there—and it will happen—some of Seattle's most vulnerable residents will be the most in need of help.

GIMME SHELTER

The quake occurred just before 3:00 P.M. local time on March 11, 2011. About eight miles off the coast of Japan, the subduction zone formed by the Pacific and Eurasian Plates—a long and often active subduction zone—gave way, resulting in an M 9.0 earthquake. That a subduction zone would be capable of an earthquake of this magnitude—the strongest in Japanese history—isn't all that surprising. As we've seen again and again, these places where continental and oceanic plates meet are capable of tremendous force.

What was distinct about this quake was just how far the ground moved in and around the epicenter—as much as 150 feet in some places. Emily Brodsky, a professor at UC Santa Cruz's Seismology Laboratory, says no one expected that fault to be so slippery. That it so displaced the earth changes what we know about the energetic potential of quakes. Another surprise was the fact that the earthquake ruptured vertically not just at its hypocenter fifteen miles below the surface, but all the way up to the seafloor, making its felt effects that much more powerful. The USGS estimates that the quake moved the main island of Japan—the entire island—by eight feet. NASA says that same force shifted the earth's axis by six or seven inches, thereby changing the planet's "wobble" as it rotates (that's their word, not mine). It also accelerated our orbit that day, shaving 1.8 microseconds off the day.

The shaking from the quake forced the shutdown of eleven nuclear reactors at four Japanese plants. At each, backup generators began providing power to keep the radioactive materials cool. Seven of those reactors were in Fukushima, a mountainous prefecture known for its proud samurai tradition, still practiced and celebrated in festivals every year, along with its national parks and farms— and its nuclear power. For years, the region had been considered a successful experiment in urban and industrial planning—a place where power plants and growers could coexist. The region was famous for its rice and peaches, but the real prize was its Wagyu cattle, known to produce excellent Kobe beef.

The 2011 earthquake changed all of that.

It's still unclear how much damage the Tohoku quake itself did to the two Fukushima power plants. Workers there told *Atlantic* reporters Jake Adelstein and David McNeill that it was significant: broken pipes, including those supplying cold water; tumbled walls; exploding oxygen tanks near the reactors. No one knows for certain just how extensive this devastation was, though, because ninety minutes after the quake the entire area was inundated with tsunami waters. They swamped the backup generators and fried the electrical systems, knocking them out of service. With them also went the plants' cooling systems. And without that, the radioactive material began to melt.

Most nuclear power plants work by splitting atoms to create heat. The heat turns water into steam, which, in turn, powers turbines not so unlike the ones we visited at Hoover Dam. Power plants control this activity in a variety of ways, including temperature. Without those cooling safeguards, the fuel continues to heat up and, left uncontrolled, will begin to release radioactive material into the atmosphere.

Just after 7:00 P.M. that night, the Japanese government declared a nuclear emergency. The prime minister ordered the mandatory evacuation of everyone living within two kilometers of the power plants. An hour later, he widened that zone to three kilometers. The next day, he enlarged it again to encompass twenty square kilometers—over 154,000 residents in all. Many left carrying suitcases and shopping bags, wearing surgical masks and hunkering down in municipal buildings and hospitals. As the evacuation area widened, they were forced to pick up and move again—sometimes as many as six times.

Andy Karam was one of the first Americans on the scene. Karam cut his teeth in radiation studies (very much a real thing) as a nuclear machinist in the navy, where he worked on both propulsion plants and fast-attack submarines. These days, he works primarily as a consultant for radiation safety and health physics. Currently, his main gig is serving as the director of radiological operations for the New York City Police Department's counterterrorism division. During big events like the Macy's Thanksgiving Day Parade or the New Year's Eve ball drop, you can find him either hovering over the event in a helicopter or skulking around garbage cans and alleys, looking for any evidence of a dirty bomb planted by terrorists. But don't panic too much about that, says Karam: "It's like what George Carlin used to say about missing car keys: You have to look in the freezer just to confirm they're not there."

He also works for NYC Medics, a nonprofit group focused on disaster relief. When the Fukushima plants went down, the Japanese government called them to help train health care providers not familiar with the effects of radiation. Karam still gets choked up talking about the disaster area, from the watermarks on the airport terminal to the teddy bears and family heirlooms half covered in muck to survivors probing mud with long poles in the hopes of finding lost family members. Equally arresting for him was the scene farther up in the mountains, where villages had avoided the earthquake and tsunami damage altogether and now stood fully intact but utterly vacant.

The mayor of one village told Karam that many of his residents didn't want to go. They were far removed from the possibility of radiation burns, and the elderly there knew they wouldn't live long enough to see any effects from more chronic conditions like thyroid cancer. They worried about their husbands and wives, not to mention the tens of thousands of cows depending upon them for their very existence. Who would feed and water them?

For most, the answer was no one. Thousands of the cows died of dehydration or starvation. Thousands more were ordered to be euthanized and buried, lest they spread radiation.

About 230 square miles surrounding the plant were evacuated in the days after the meltdown began. Just a year later, the estimated economic loss to that region was in the neighborhood of US $500 billion. Cleanup will include dealing with hundreds of tons of contaminated coolant water and 22 million square meters of contaminated soil: roughly thirty-four Disneylands' worth (high-pitched mouse and his girlfriend included).

※

It's easy to assume Fukushima was a freak event. In general, nuclear power plants tend to perform really well in earthquakes. In the case of older plants, some of that success can be attributed to little more than dumb luck. John Stamatakos serves as the director of the Center for Nuclear Waste Regulatory Analyses Department at the Southwest Research Institute, where he performs studies on behalf of the Nuclear Regulatory Commission. Historically, he says, seismic considerations for these plants were based on fairly simple hazard calculations: "We'd try to figure out the biggest possible earthquake on the nearest fault source to the plant, then calculate what ground motion could do at that plant and design for that."

The problem, he says, is that earthquakes have a way of surprising us. Take the 2011 Virginia quake. Its magnitude exceeded this hypothetical extreme case (known in the industry as an OBE, or operating basis earthquake) and shut down Dominion's North Anna Power Station for a couple of months while contractors assessed damage and Dominion sought to prove that the plant was in good working order. That's NRC regulation for any OBE event, and 2011 wasn't the first time that policy was needed. In 1979, construction on the Virgil C. Summer Nuclear Station in South Carolina was nearly complete when a series of earthquakes prompted by the filling of its adjacent reservoir exceeded that plant's OBE. Seven years later, an M 4.9 earthquake rattled northeastern Ohio, home of the newly built Perry Nuclear Power Plant, which was still awaiting its certification. As early as 1988, researchers found evidence that that quake was caused by injection wells nearby.

Naturally caused seismicity poses no less of a problem. In 2016, California reluctantly ordered the closure of Pacific Gas and Electric's Diablo Canyon plant after previously unknown faults were discovered in close proximity to the facility. After the 2011 Virginia quake, reporters at the *Richmond Times-Dispatch* uncovered internal memos at the Justice Department and NRC indicating that they knew about faults near the North Anna plant but said nothing.

By all accounts, the 2011 quake did little damage to North Anna: a few tiny cracks, but none on load-bearing walls. It did rattle casks containing spent fuel rods, moving about half of the fifty-three 115-ton casks up to four inches on their concrete pads. The damage could have been worse: After the Fukushima disaster, the NRC ordered a new seismic study of all existing nuclear power plants. At North Anna, they found fire-suppression systems not built to seismic design. Plus, they said, one of the plant's turbine buildings was missing a key flood wall.

North Anna was one of at least five plants cited by the NRC as vulnerable to seismic activity. Headlining that list is a real doozy in the nuclear world: Entergy's Indian Point Energy Center, located just twenty-four miles upriver from New York City.

I first visited Indian Point on a rainy morning in February. The day before, Punxsutawney Phil had emerged to an absence of his shadow, leading his handlers to proclaim we were all in store for an early spring. Little evidence of his meteorological prognostications could be found in the town of Cortlandt, New York, where a group of activists was holding a vigil outside the tiny courthouse. Inside, nine protesters—including a baseball-cap-wearing Episcopal priest— were awaiting their first appearance after attempting to block the construction of a natural gas pipeline alongside the power plant.

Serving as an expert witness on their behalf was Paul Blanch, a longtime nuclear energy consultant. Blanch spent decades working as the safety manager at plants including Connecticut's Millstone Nuclear Power Station and Indian Point Energy Center. He served under Andrew Cuomo as New York State's chief nuclear consultant and expert witness. In 1993, he was asked to testify before Congress about the importance of protecting whistle-blowers at the NRC. In other words, he knows his way around a day in court. He'd suggested that I keep him company on the bench outside the courtroom while he waited to be called—it'd be a good opportunity to talk about seismic design and safety, he said, and he was certain we'd have lots and lots of time.

With his navy suit, matching trench coat, and pack of cigarettes, Blanch is straight out of central casting for *Mad Men* or another period drama. It's hard to not get overly excited about the prospect of conspiracy when he talks, since everything he says through his smoker's croak sounds like it came from Deep Throat himself. We didn't get very far in our conversation when the judge in the case announced he had to recuse himself—apparently, he had served on a committee with one of the defendants. And so Blanch and I decided to brave the freezing rain to visit the notorious power plant itself. We got there in his brand-new Corvette (bloodred), and as we drove, he switched over to a vapor cigarette so that I wouldn't be bothered by the smoke.

Now retired, Blanch spends most of his waking hours trying to call attention to the dangers in our crumbling nuclear infrastructure. He says he worries most about Indian Point. As we zip around in his sports car, Blanch lists some of the safety issues to befall the plant in recent years: contaminated water leaks, structural weaknesses in its reactor domes, transformer fires, power failures, falsified records, and a host of unplanned shutdowns. There's an aged gas pipeline that runs pretty close to directly under the plant; some authorities, says Blanch, believe it was one of the potential targets for the 9/11 attacks. This new pipeline, he says, has the potential to be even more dangerous. He's taken to calling the possible resulting disaster from a pipeline explosion in the vicinity of Indian Point "Fukushima on the Hudson." A lot of media organizations have picked up on that idea.

No one gave the prospect of an earthquake here much attention until the USGS changed its seismic hazard maps a few years ago with new attention paid to the Ramapo Fault. The NRC also adjusted its rankings of power plants at seismic risk. They put Indian Point at the top of that list.

Klaus Jacob, special research scientist at Columbia's Lamont-Doherty Earth Observatory, says this kind of attention is long overdue. Unlike North Anna, Indian Point is located pretty much on top of a fault zone. Even if it weren't, Jacob cautions, it would be unwise to compare the two.

"North Anna was built much later and, from a design point, seems to be aging much better," says Jacob. "Indian Point was one of the earlier nuclear power plants. The industry has learned a lot since then."

He also points out that not all earthquakes are created equally: Magnitude and distance matter, yes, but so, too, does the frequency of the shaking and the type of soil and rock through which the waves are passing. Indian Point, he says, sits on particularly hard, dense rock, which won't dissipate wave energy like, say, the sandstone of the West or even the soft fill layered below much of Washington, DC. A magnitude 5.0 quake, he thinks, would be enough to raise safety concerns at Indian Point, particularly with internal structural components like relays and bolts—both of which are particularly susceptible to damage from high-frequency motion. A 6.0 within five to seven miles of the power plant, he says, would be enough to endanger the domes containing the reactors themselves.

He and Blanch both agree that so much of the plant's response to the earthquake has to do with conditions inside. Take Ohio's Davis-Besse Nuclear Power Station. Everything seemed aboveboard there until a routine refueling outage and inspection. As inspectors looked over the reactor pressure vessel—the housing that contains the core and its rods—they found significant cracking in several of the nozzles there. Further inspection revealed that these cracks had been leaking boric acid, which in turn had been slowly eating away at the lid of the reactor. By the time it was discovered, the acid had eaten a patch of the lid to a thickness of less than three-eighths of an inch. Had the reactor been jostled in a quake, that force may well have been enough to break through the remaining liner.

Davis-Besse was an unusual case, but not an entirely isolated one. The year 2015 witnessed ten "near misses" at nuclear power plants around the country (near misses are defined by the NRC as events that potentially increase the risk of a reactor meltdown by at least ten times). Half of those near misses occurred at plants owned by Entergy, which runs Indian Point. They include malfunctioning pressure valves at Massachusetts's Pilgrim Nuclear Power Plant after a winter storm knocked out power to the plant and pooling water caused by the fire protection system at Indian Point. That may not seem like a big deal, but it turned out to be a far greater concern than the transformer explosion that initiated the sprinkler systems. According to the Union of Concerned Scientists, once that water reached five inches, it would have short-circuited the electrical system, causing a total blackout at the plant. That same year, the NRC also responded to security issues at North Anna and reports that workers had accidentally drilled through crucial metal reinforcement within the concrete surrounding the reactor at Virgil C. Summer.

By all accounts, a seismic event capable of creating a nuclear meltdown in this country is unlikely. But, says Jacob, that doesn't mean it's not worth planning for. It all goes back to that idea of risk: Even if the chance of an event occurring is really low, its potential impact may well make it worth our time to anticipate.

Megan Pribram is assistant commissioner of planning and preparedness at the New York City Office of Emergency Management. She says their message in the event of a disaster at Indian Point would be for residents, at least initially, to shelter in place while state and city health officers worked with the federal government to determine if an area needed to be evacuated.

In the event that an area did need to be evacuated, the city's police department would take the lead in making that happen. Pribram wouldn't go into any detail about what these evacuation plans might look like or what kind of event would prompt them.

"We want people to tune in and listen to the message before acting," she says. "And we want our plans to remain as flexible as possible."

That seems just fine to Andy Karam. He says the evacuations in and around Fukushima came with their own casualties—particularly among the most vulnerable populations there. Around 2500 evacuees spent two weeks in the Saitama Super Arena (think Superdome); another 1700 were moved multiple times to different schools in the region. These populations had probable PTSD rates upward of 67 percent.

For the elderly and infirm, these mandatory evacuations were difficult transitions. Most were moved by bus or police vehicle and did not have attending medical care. Some went forty-eight hours without food or water. A study in the scholarly journal *Critical Oncology* found that as many as fifty elderly nursing home residents died of either dehydration or hypothermia during that time. Thousands of additional patients had to be evacuated from hospitals and assisted living centers outside the evacuation zone because delivery workers would not bring needed supplies for fear of radiation poisoning.

The World Health Organization estimates that over 1600 people died as a result of this evacuation—more people than were killed by the initiating quake (though far less, of course, than the 20,000 killed in the tsunami).

Karam thinks a lot of those deaths were avoidable.

"When the reactors went down, the Japanese government assumed the worst. I can't fault them for that," he says, "but it was a panicked response that cost lives."

Meltdowns are not the same kinds of events as, say, dirty bombs or nuclear missiles: The radiation from them takes time to increase, and they are rarely accompanied by acute conditions like radiation burns or poisoning. Instead, he says, it's the gradual accumulation of radiation from meltdowns that more commonly leads to chronic conditions like thyroid cancer. With that in mind, he thinks a more prudent approach would be to set targets—a radiation level at which it's time to prep people for an evacuation, another for when it's best to start moving able-bodied people, a higher one for the sick and infirm.

"Evacuation comes with all kinds of negative consequences," he says, "even among the physically healthy."

He points to studies completed after the Chernobyl disaster. The World Health Organization found that the very process of evacuation proved what it called a "deeply traumatic experience" to many evacuees, for a variety of reasons, including what the WHO calls "disruption to social networks," the realization that one can't return to his or her home, and "a social stigma associated with being an 'exposed person.'"

As a result, evacuees have suffered from much higher rates of mental health issues than those who were not evacuated.

An early study on the effects of the Fukushima disaster came to many of the same conclusions. Evacuated residents there have experienced far higher rates of PTSD than other earthquake survivors in the country. Suicide rates have remained demonstrably higher among that population than among other earthquake and tsunami survivors, too.

That's one big reason Karam agrees with Pribram: "Unless there's a really extreme circumstance, I tell people to hunker down at home."

Radiation dose rates, he says, fall considerably inside buildings—especially high-rise apartments and hospitals (though he recommends you stay away from higher floors and windows). Even moving away from an external wall and toward, say, your dining room table can make a big difference, too. Just don't drink the water, he cautions: There *is* a pretty good chance it may have unsafe radiation levels.

"Make yourself comfortable, open some canned food, and live off the juice and beer in your fridge," Karam recommends. "Probably you'll be fine."

—————

Of course, mental health effects aren't limited to evacuees. And there seems to be something particularly affecting about the experience of surviving the trauma of an earthquake. In her book *The Earthquake Observers*, historian Deborah Coen traces centuries of psychological responses to earthquakes. Again and again, doctors found that earthquake survivors presented with particularly dramatic cases of traumatic shock and other mental ailments. Nineteenth-century French neurologist Jean-Martin Charcot observed that no other disaster, natural or man-made, had the ability to evoke "stronger emotions, more intense fright" than an earthquake. Other early psychologists found that quakes tended to turn people into "hysterics and liars" driven by "nervous excitability."

This is no mere Victorian hyperbole. Over a third of the survivors of the 1994 Northridge earthquake found themselves confronted with an "emotional disorder" warranting treatment. In the New York and Los Angeles scenarios, that would include 2.5 and 8 million people, respectively. Following the 1999 Taiwan earthquake, health care providers found a disturbing spike in suicides among survivors.

Psychologists studying survivors of the 2005 Kashmir earthquake found that those individuals had considerably higher rates of psychosis, including cases of paranoia and hallucinations. Following the New Zealand quake, mental health hotlines reported huge spikes in calls, particularly from parents whose children were experiencing panic attacks. Survivors of the recent quakes in Italy reported experiencing "phantom quakes" in which the brain tricks the body into believing it is experiencing more tremors. For some survivors, this went on for months.

And while we'd like to believe that this kind of trauma can have a galvanizing effect on a community, history suggests the opposite is often true. Susan Millar Williams and Stephen Hoffius spent years researching their account of the aftermath of the 1886 Charleston earthquake, which sparked racial tensions and violence, including the murder of one community leader. While they were working on the book, they also observed our collective behavior in the wake of 9/11, Hurricane Katrina, and the Haitian earthquake. Yes, there were great examples of heroism and unified communities, say Williams and Hoffius, but also dramatic examples of resource hoarding and adversarial stances.

"Natural disaster stresses everyone. If you're not in very close control of your emotions anyway, anything that's raw is going to come to the surface as anger and frustration," says Williams. "We want to romanticize disasters as an occasion of unity. But in many cases, people already feel like there's not enough to go around, that there will never be enough to go around."

She says she experienced that firsthand in the days after Hurricane Matthew, when even residents in her middle-class neighborhood began eyeing one another with suspicion at donation centers offering clothing and home supplies.

"Truck after truck would pull up with sofas, mattresses, Ralph Lauren pink and tan argyle socks, Russell athletic wear, and more. Stuff just poured into this community: It was more than enough for everyone to have their share. And yet this great bitterness emerged that someone was getting more than you."

That bitterness can cause the breakdown of the social contract. Multiple studies have found that crime increases after natural disasters: not just offenses you'd expect like robbery, but also violent crimes like rape and murder. Haiti experienced a significant spike in gang-related crime after its 2010 earthquake; New Zealand experienced an increase in domestic violence and sex crimes after its quake. The World Health Organization reports similar upticks in violence ranging from elder abuse to child abuse after earthquakes, and they say the disruption quakes cause also makes it easier for human traffickers to succeed. In the days after the 2015 Nepal earthquake, human rights workers intercepted at the border over seventy young girls, all of whom had been kidnapped.

Why? Because, write World Health Organization researchers, "disasters disrupt the physical and social environments that shape health." With that disruption comes the breakdown of some of our most sacred social contracts. And so one of the most important questions we all need to ask when faced with the next quake is this: Who do we want to be in its aftermath?

THE SCHOOL

I t's not yet 6:00 A.M., and I am waiting in the parking lot of a bakery still closed to customers. Inside, the early shift is arranging trays of premade muffins and Danishes, lugging out carafes of coffee and cream. A few minutes later, Brian Atwater and David Yamaguchi show up in Atwater's very well-used Honda Accord with his even more well-used canoe strapped down to the bumper. The car is overflowing with shopping bags of books and paddles and a very big hunk of wood. The two have a well-rehearsed routine with each other that is one part Abbott and Costello, one part *Odd Couple*. They've known each other for thirty years now; Atwater's two-year-old granddaughter calls Yamaguchi "Uncle David."

Atwater is scientist emeritus for the USGS based at the University of Washington. Yamaguchi is a dendrochronologist—someone who studies ancient trees to make sense of the world hundreds or even thousands of years ago. Together, they've revolutionized everything we know about the history of enormous earthquakes in the Pacific Northwest. Today, we are heading to the Copalis ghost forest so that they can show me how.

The three of us pop into the now-open bakery for coffee but buy nothing to eat: The highlight of the trip for both of them is a crunchy bakery located outside Olympia, where the pastries are much, much better, Yamaguchi says in a whisper so the staff here won't hear. Inside the car, Atwater offers me a well-used plastic container that once housed black-bean-and-corn salsa and now contains raspberries from his garden. Yamaguchi squeezes into the backseat, re-arranging the books and life jackets and things so there is enough room for his thin frame. We're on our way to the base of the Olympic Peninsula, a 150-minute drive (longer if you stop at the bakery, which is fabulous, especially if you like gingersnaps). As we drive out of Seattle proper, Yamaguchi points out the metal jackets on over-pass pillars intended to keep the concrete from shattering in a quake.

"You have Brian Atwater to thank for that," he says. Atwater just shakes his head.

We pass salt marshes and tidal flats that were once forests, before the land dropped suddenly and violently and became something else. Atwater says those changes were caused largely by tsunamis and can still be traced on the land. I want to know how Atwater even thought to look.

As a graduate student in the 1970s, Atwater had been tasked with what seemed like the least sexy of research projects: studying the remnants of core samples taken from the Sacramento and San Joaquin Rivers before construction of some of the bridges across the bay. Engineers needed to know how to anchor those bridges in a mucky environments with dramatic tides and even more dramatic ground shaking, and the samples were intended to tell them how. That had been years ago. In the interim, those samples had lan-guished in normal kitchen Ball jars sealed with standard metal lids as if they were pickles or strawberry jam made by your grand-mother. Some had been sitting around for over thirty years, drying out and getting dusty and generally being ignored. Atwater was in-terning at the USGS, and one of his supervisors thought a really great project for graduate student interns would be to examine and catalog the contents of those jars. So Atwater and another budding geologist or two spent their time in an even dustier warehouse picking through what he calls "cast-off" bridge cores and generally wondering if this was, in fact, the right profession for him.

As it turns out, it was. These Ball jars contained layer after layer of sediment taken from tidal estuaries. Some were studded with what Atwater calls "really beautiful gypsum crystal" (he is a geologist, after all), but most were just strata of varying colors of brown, along with plant matter in varying stages of decomposition. Atwater didn't know a lot about plants at the time—especially about tidal plants—but he quickly learned. And as he did, he also discovered that you can tell a lot about the history of a place based on what is in its mud. He found pickle weed and seaweed and sphagnum and tiny bivalves, all proof of the succession of the bay as it ebbed and flowed. At times, dry land had stood there. Other times, it had been marsh. Still other times, as it is now, it stood fully submerged.

"You didn't need a microscope or any special technology to see the significance there," says Atwater. "The ocean had written its history right there in a way anyone could observe."

The more he studied, the more Atwater realized he liked the work and the work liked him. The USGS asked him to stay on and stationed him not in a warehouse but, instead, in the San Francisco Bay. It was a perfect place for his kind of work: Twice a day, tides rushed out to reveal striated riverbanks, each line a chapter in the area's deep history. He became an expert in color nuance: Lines of blue-gray meant clay deposits; rich chocolate browns indicated river sediment; sienna and sepia, decomposing grasses.

In time, the USGS moved him and his family to the Seattle area—just before the 1985 Mexico City quake struck. From a geological perspective, it was a quake that wasn't supposed to happen—at least, not the way it did. Seismologists knew about the massive ruptures caused by subduction zones. They'd studied the 1960 quake in Chile, the 1964 Alaska quake and its aftershocks. But those were different. They'd wreaked the worst of their damage on shoreline communities and habitat. The 1985 quake had all but leveled a major city—and one hundreds of miles from the earthquake source.

The seismology community, Atwater says, was "haunted by the specter" of that quake and what it meant for urban areas. So, too, was the Department of Energy. They asked Tom Heaton to conduct a study of the Cascadia Subduction Zone to see if it might rupture like the plate in Mexico had. They were interested in expanding their nuclear facilities in the region as part of a project for the Washington Public Power Supply System, or WPPSS (which is pronounced "Whoops," making it worse than WIPP and quite possibly the worst acronym in the history of the world). As early as 1970, seismologists had confirmed the presence of the Cascadia Subduction Zone, but with no evidence that it had ever ruptured, they had surmised that it was aseismic or, at best, just slowly creeping along.

Heaton cut his teeth studying subduction zones in Alaska for the oil industry. He thought Cascadia looked a lot like the Alaska zone—and similar subduction zones in Chile, too. Probably, then, it, too, was capable of a big one.

Heaton's claims were controversial in the geology world. There was absolutely no evidence of any seismic activity to support them. Today, says Atwater, the debate over how much the fault zone moves would be settled pretty easily with some GPS work, which could mark the precise location of the plates and record even the tiniest movements. That wasn't an option then, says Atwater. "If it were, geologists wouldn't have had much of a contribution to make."

And so the debate continued. Atwater attended one of Heaton's talks on the subject and was immediately intrigued. More than that, he thought he might be able to help.

Atwater knew the oral history of First Nation and Native American peoples in the area included stories about what sounded a whole lot like a megathrust quake. The Yurok people told stories of when their ancestors had run to the top of a mountain wearing bird feathers to escape a flood of water. The Quileute told stories of how Whale and Thunderbird once had a fight so terrible that it shook the ground and knocked over trees. Every tribe, it seemed, had some variation on this story: of awful shaking and shattered villages and people being swept away.

But was it caused by a subduction quake? And if so, when?

Atwater figured that if such a quake had happened, two very visible things would have occurred: The ground would have subsided, and there would have been a tsunami. Both would have left a mark. He'd been to Chile and Alaska and seen the devastation there, the way the estuaries told the story of a collapse only a megathrust could create. "Earthquakes like that write their own history," says Atwater, "and often in surprising ways."

In Chile, earthquakes had created massive uplift in places, raising what had been beach or underwater coral reefs tens of feet in the air. Entire Japanese forests had been obliterated by tsunamis, leaving wetlands and root systems where enormous pines once stood. That 1964 Alaska earthquake created mudflats where towns once stood. It raised places like Middleton Island by about twelve feet and created a new marshy intertidal zone around the entire island.

Atwater decided to search for that kind of evidence on the coast of Washington, too. He began his search on the Neah Bay, part of the Makah's tribal lands. There, he found what he calls "a beautiful stack of peat recording the rise of the sea" (he is a paleobotanist, after all). He also found evidence that the ground had dropped suddenly and dramatically. It was much like the Hebgen campground but far more dramatic. Something really enormous had done this. But when? Atwater kept searching estuaries up and down the coast and found remarkably similar evidence. In between the peaty layers of mud and sediment, they also found sand. That, they surmised, might very well have been brought inland by a major tsunami.

It was raining particularly hard that day. Atwater stopped into a general store to take a break and warm up. A rack of postcards stood near the cash register. He bought one and sent it to Heaton: Here, he figured, was important evidence to support his theory. He figured Heaton would want to know.

Atwater had enough evidence to warrant the postage stamp, but nowhere near enough to persuade the scientific community. He still didn't know for sure what had killed these trees and when. Those Ball jar samples from his grad school days showed evidence of gradual sea level rise and fall. Atwater couldn't say for sure that the same thing hadn't happened here. He had evidence of a dramatic tsunami but couldn't say when it had occurred.

That's where David Yamaguchi came in. Up and down the peninsula, Atwater had found stump after stump of Sitka spruce and red cedar submerged well below high tide. Something had killed these trees. If he could figure out when they died, he could determine whether a megathrust quake was responsible.

Yamaguchi was a post-doctorate at the time. He'd been studying ancient tree rings to chart volcanic activity at Mount Saint Helens. Atwater thought Yamaguchi might be able to do the same for seismic activity on the coast.

Dendrochronology is a painstaking science. Basically, you take the trunk of a tree whose age you can verify and you count the rings. In this case, you count hundreds and hundreds of rings. Some are wider than others—they demonstrate seasons of dormancy and growth, years of rain and drought—with tiny variations that distinguish one year from the next if you know how to look. Yamaguchi could tell Atwater a lot about the lives of these trees if he had a full trunk. He could tell, for instance, whether the tree had gotten sick and struggled to survive as salt water gradually overtook its forest. Or he could say whether something sudden had killed those trees.

For this to work, they needed a full tree trunk. And that proved difficult to find. When submerged, spruce begins to decompose quickly.

A lot of the stumps Atwater found were missing too many outer layers for Yamaguchi to date them. The two spent a very rainy season mucking around estuaries and swamps looking for better samples. Atwater scanned aerial photographs and topographical maps looking for better sites. Eventually, he found the Copalis River. And then he called Yamaguchi.

There, a short paddle up the tidal waters, they found a haunting stand of dead cedars—ancient, enormous trees bleached by seasons of storms, their hollowed-out trunks splintered into jagged geometries. A ghost forest. Yamaguchi saw in them the missing piece of the landscape puzzle. Unlike spruce, cedar is naturally rot-resistant. And while the upper reaches of the cedars had become splintered husks, their stumps remained perfectly intact.

"This could actually work," Yamaguchi says he remembers thinking when he first spied the stand.

That stand was the destination for our canoe trip. Atwater, who by this point was wearing floppy waders and an even floppier hat, would hop out periodically to show me the layers of sediment and sand and tree root that proved the ground had subsided. In time, I came to understand and even identify the evidence of subsidence, but something about it remained theoretical. It was ultimately just layers of different shades of brown. The ghost forest is different. Something undeniably dramatic happened there. One day, massive trees grew strong and shady. And then, almost en masse, they died. What remains are weathered spires on a broken plain. They are a ghost forest the way old mining camps are ghost towns: stark in what remains, and even starker in what is missing.

After Atwater and Yamaguchi first discovered this place, the two scientists spent the summer taking tree samples from the ghost forest. In a stroke of luck, they found stumps of similarly aged trees farther inland. Atwater and Yamaguchi sent the samples off for radiocarbon dating. It is, as it turns out, a frustratingly imprecise tool in cases such as this one. The duo had confirmed that the trees died sometime between 1695 and 1720, but that was all. It still didn't answer whether one massive quake had killed these trees or whether a series of large quakes, like the ones in New Madrid, had done the job.

And so their research stalled. A couple of years went by. And then they met a Japanese historian who had been studying records kept during the time of the shogun. To the last, they were painstakingly exact. And like many of the oral histories from the Pacific Northwest tribes, they told stories of flooding and devastation: a nighttime flood in 1700 that had destroyed homes and forced people to higher ground. Its reach had been significant—nearly as much so as the 1960 tsunami created by recorded history's strongest quake. But these records never mentioned an earthquake. Atwater and Yamaguchi termed it the "orphan tsunami." The name stuck.

That, as it turns out, was great news for Atwater and Yamaguchi. They had a quake with no clear sign of a tsunami. Japan had a tsunami with no clear sign of a quake. Even better, Japan had a precise date: 1700, well within the range established by Yamaguchi's carbon dating.

The combined work of this project proved that Cascadia was capable of major thrust quakes—that one maybe as strong as 9.0 had struck there just 300 years prior.

—⁓—

In the past fifteen years, just a handful of tsunamis have been responsible for the deaths of hundreds of thousands of people. They've destroyed entire communities and rendered entire swaths of land uninhabitable.

In some regards, tsunamis are a lot like regular ocean waves: They're created by disturbances; they have crests and troughs; they are measured by their height and length. But there are important differences that make these phenomena so deadly. For starters, it takes a really, really big disturbance to make a tsunami. Often-times, that's a subduction zone quake. But not always. Enormous underwater landslides, volcanic eruptions, and even meteor strikes can cause them, too. What all of these stimuli have in common is that they create an abrupt or sudden shift that displaces water. Drop a long log in a pond: You'll see the way its contact forces water out from either side of it, not unlike what happens when you throw a pebble into that same pond. The difference, though, is that instead of rings of disturbances, the log sends out a series of more linear ones, also known as trains. Tsunamis are like that. An initiating event like an earthquake or a landslide pushes two waves of water out from the disturbance site—one traveling toward shore, the other traveling farther out to sea.

Waves travel exceedingly fast over open ocean since there's little resistance to slow them down. They've been clocked moving at speeds of 600 miles an hour, which is about the same speed as a commercial jet. Because these waves meet such little resistance in the open ocean, their energy is disbursed such that they can seem hardly noticeable to boaters there. Japanese fishermen named them "*tsu nami*"—"harbor wave"—as a way of trying to make sense of this phenomenon: They'd spend their days at sea, where nothing seemed amiss, only to return home and find their port and village destroyed.

That same phenomenon is what so delights surfers around the globe. As soon as a wave encounters resistance in the form of a coral reef or rising seafloor, it can no longer diffuse its energy downward. The front edge of the wave slows down considerably, causing the rest of the wave to stack up on top of it, eventually toppling over itself.

Survivors of the 2004 Indian Ocean tsunami describe those waves as walls of black water—so black, in fact, that multiple people thought it was oil and not seawater. They recall the wind those walls of water produced, the way they seemed to rise sixty feet or higher, the sheer solid mass of them, the deafening roar of a freight train as they traveled at speeds of fifty miles an hour or higher and devoured everything in their path. Whether they were British tourists or twelve-year-old residents, whether they happened to be in Sri Lanka or Thailand, those survivor accounts return again and again to that same description: a black wall as solid as any rock or metal, unrelenting and moving at terrifying speeds. Like something straight out of a Hollywood movie.

It takes an extreme earthquake to create that kind of tsunami wave. Most of the time, they appear more like a superaccelerated tide—a surge or a flood—than a wall. Still, the average tsunami is no less deadly. The largest tsunami wave ever recorded was in Lituya Bay, Alaska, in 1958. By subduction zone standards, the earthquake precipitating it wasn't all that huge—an M 8.3—but it created a wave estimated to be a hundred feet high. More importantly, it pushed water 1700 feet above sea level. This height is known as the run-up of the tsunami, and it's that run-up more than the height of a single wave itself that does the most damage.

Tsunamis often come in multiple waves, and it's not uncommon for the highest of these to be several waves back. This counterintuitive fact has resulted in a lot of deaths: People think the danger has passed with the first wave, only to have an even more massive one bear down on them. That's what happened when a tsunami prompted by the 1960 Chilean quake reached the town of Hilo, Hawaii. Warning sirens there began sounding around 8:30 P.M. that night. Just after midnight, the first wave arrived. It was just a few feet high, and most residents assumed the danger had passed. Some returned to their homes. Others had never left. An hour later, remaining residents were awakened by the sound of growing thunder. Within seconds, a surge of water inundated the town, knocking homes off their foundations and sucking cars out to sea. The total run-up of that tsunami wave was thirty-five feet. Sixty-one people died that night. Close to 300 were injured.

We tend to think of tsunamis as a Pacific Ocean phenomenon, and that's a dangerous proposition, says Atwater. His current research is looking for evidence of the waves on the Caribbean island of Anegada, a low coral island, the northernmost of the British Virgin Islands (this, by the way, is a very plum assignment if sand beaches and aquamarine water are your thing). There's evidence of about fifty tsunamis that have made it that far since 1530, including one prompted by the 1755 Lisbon quake. Believed to have measured M 9.0, the quake sent three tsunamis barreling down on the Portuguese city. They also hit the Azores and the Canary Islands as well as the Caribbean, where waves nearing fifteen feet were recorded. Thirty-foot waves were recorded in Guadeloupe after the 1867 quake. But perhaps most damaging was the 1918 earthquake that ruptured off the coast of Puerto Rico. Its subsequent tsunami sent twenty-foot waves to that island's coast, killing forty people. They also made it as far as New Jersey, though they did little damage. About 20,000 years ago, a massive underwater landslide prompted a tsunami that hit Virginia and North Carolina, and there is even evidence of tsunamis as far north as Maine and the coastal Maritimes.

The chance of a major event on the Eastern Seaboard is slight. Some geologists think a volcano on the Canary Islands could create a massive tsunami as high as seventy feet, but that would take an enormous eruption followed by the landslide of the volcano's entire flank—a highly unlikely scenario. When both Krakatau and Santorini last erupted, they created waves in their area but nothing like the kind of geologic displacement a tsunami of that size would need. Researchers at the Lamont-Doherty Earth Observatory put the chance of an East Coast tsunami in any given year at about 1 in 1000. That number means very different things to different people and communities. It's the same chance that a healthy forty-year-old male will die of stroke or colon cancer or that a contestant will win the jackpot on one of several major game shows or that asteroid 1999 RQ36 will strike the earth in 2182 according to Spanish astronomers (NASA puts the chance of that event at more like 1 in 3500).

For communities like Myrtle Beach, South Carolina, and Norfolk, Virginia, that risk, however seemingly slight, has been enough to prompt action. They received "Tsunami Ready" certification from the National Weather Service by training civic employees, creating a tsunami emergency plan, ensuring that residents can receive tsunami warnings in multiple ways, and installing signage that both alerts people to the potential danger and includes evacuation routes. At last count, only about twenty communities on the Eastern Seaboard have participated.

—m—

Aberdeen, Washington, is a working-class town defined by a busy port at one end and a massive timber mill at the other. In between these two bookends sit mostly small one-story homes set on small lawns. The blocks are perfectly symmetrical and very, very flat. It's easy to confuse one for another, with the possible exception of the one containing the childhood home of Kurt Cobain, the iconic lead singer of Nirvana who inspired a generation of flannel-wearing grunge lovers (not to be confused with flannel-wearing seismologists, who are a different breed entirely). Cobain died in Seattle, but he spent his youth brooding along the Chehalis River in Aberdeen. There's a strange little memorial to him on the waterfront there that is oft visited, and you can even sit under the bridge where he did most of his ruminating, though I wouldn't suggest it unless you've had a hepatitis shot and aren't afraid of dirty needles.

Farther inside the town sits a wide main street with a wholesale fruit and vegetable market (Brian Atwater and David Yamaguchi recommend the cherries there), a couple of banks and car dealerships, and a lot of signs warning about tsunami evacuation routes.

I meet Chuck Wallace, who oversees emergency management in the area, under one such sign. He'd suggested that we have lunch at an Italian restaurant also near that sign, which has a veal scampi that reminds him of his native Philadelphia. If you were going to cast a remake of *The Sopranos*, Wallace would make a pretty good candidate. He has a shock of white hair, a love of nice shirts, and the kind of bravado that would fit right in.

When I order a salad, he laughs at me, which I tell him also seems like a native Philadelphian thing to do. In an attempt to redeem myself, I tell him I'll order it with grilled salmon.

"Suit yourself," he says with a shrug.

Wallace worked for years as a fire captain back in Philly. That was originally the thrust of his job in Grays Harbor County, Washington, too. He thought of it as a kind of staged retirement, even though it came with the usual associated duties of managing various emergencies.

But these days, tsunamis are absolutely his specialty. Recently, a group of geographers and geologists conducted a study of the Cascadia tsunami zone, which stretches from Washington State down into Northern California. About 95,000 people live in this zone. Nearly half of them—about 48 percent—live in the counties for which Wallace is responsible. He takes this responsibility very seriously.

It's a niche market. North America has had even less experience with tsunamis than it has with earthquakes. There was the tsunami prompted by the 1960 Chilean quake and the Alaskan event in 1964. In 1929, an earthquake off the Grand Banks caused a pretty good one in Newfoundland. Survivors remember a black wave that first tore through town and knocked homes off their foundations and boats off their moorings, then sucked them back to sea. Dramatic photos of the surviving homes still remain. They're being towed to shore by schooners.

But these few events are not a lot to hang your hat on if you're a town official or, even worse, a resident wondering if you should even be concerned. That's where Wallace comes in. Geological modeling says the chance that the entire Cascadian fault will rupture like it did in 1700 is pretty slight—about 5 percent, by most counts. More likely, they say, just one or two of the segments will go. That'd probably result in an M 8.5 quake.

In either scenario, there's a good chance the earthquake would result in a tsunami (though its height and power would be hugely impacted by the magnitude). Tsunamis come in two varieties. Think back to the log analogy. When you drop it into the pond, water radiates from either side. That's true in the case of subduction zone quakes, too. So what you tend to get are actually two sets of tsunamis traveling in opposite directions from each other. One is called the local tsunami (that's the one that hit Cascadia in 1700); the other is the distant tsunami (that's the one that made it to Japan). When the 1960 Chile quake ruptured, it took nine hours for the subsequent tsunami to hit Hawaii—plenty of time, at least in theory, to warn residents and begin a methodical evacuation. A local tsunami can inundate a region in twenty minutes or less.

The National Oceanic and Atmospheric Administration maintains the National Tsunami Warning Center. It used to be called the West Coast/Alaska Tsunami Warning Center, but NOAA changed it after worries that residents of other coastal states would think no one was looking out for them. Today, the center monitors the Atlantic and Pacific Oceans by way of a series of detection buoys and tidal gauges. It's part of the National Tsunami Hazard Mitigation Program, which in turn is part of the International Tsunami Information Center based in Hawaii. They're really good at detecting distant tsunamis. They can usually register local ones, too, but not necessarily in time to issue a warning.

That creates challenges for emergency managers. Further complicating the matter is the fact that a local tsunami also comes with local (and often severe) earthquake damage. And so, in the case of a local tsunami, the only warning coastal residents in the Northwest may get is the ground shaking beneath their feet. If it's bad, their routes to higher ground are going to be limited.

"It's really hard for the human mind to grasp the damage created by a 9.0 quake. After a 9.0 you're not driving anywhere," says Wallace. "You're walking over debris—if you're lucky."

The communities in and around Aberdeen have a large elderly population. They also have a disproportionately high number of single-mother households with small children (these homes make up about 30 percent of coastal Washington, compared to just 5 percent in other parts of the inundation zone). Were a tsunami to hit, they'd have just twenty minutes to reach higher ground. That same study I mentioned earlier estimates that about 15,000 of the residents in coastal Washington might not make it in time. That's particularly true for the young kids and older folks. Wallace spends a lot of time thinking about them. And about the inmates at a juvenile detention center located just outside of town. As it is, some of the supposedly secure doors—the ones intended to confine residents—don't close all the way because of the last quake. He thinks about the parents of those kids a lot. "We tell them we're going to take care of those kids," he says. "Are we?"

For Wallace, the likelihood of the disaster doesn't matter. What counts is how to minimize it. He gets angry at doomsday predictions that the area is going to be obliterated by a tsunami. It doesn't have to be that way. And he'd like to believe he's made real progress in creating a different ending to the story.

Recently, the town implemented a "Yellow Brick Road" program to direct people to higher ground in the area. Wallace persuaded the police chief to put the signs in Spanish, too. The official position of the county is that its population is overwhelmingly white and English-speaking. Hispanics, according to those figures, make up a very slim minority—about 4 percent. Wallace takes issue with those numbers.

"Spend an hour at Walmart and you'll eclipse that number a bunch of times over," says Wallace.

A lot of these immigrants work on the crab boats. Wallace has a pretty good idea they may not be here legally. They go out of their way not to be counted. People in city uniforms make them nervous.

One Sunday, he visited a Catholic church that plays home to a large number of migrant workers; most of the parishioners split after Communion rather than wait around and figure out why he was there. The church houses a large homeless population, too. By law, they can stay in the church for up to ninety days. The people who avail themselves of that tend not to want to talk to city officials, either.

"It scares me when I can't reach them," he says.

Wallace has taken to hiring kids out of the high school Spanish classes. They're less intimidating, and they can explain why it's really important to read the flyers Wallace had produced. He regularly stops tourists to the area and makes sure they know the drill.

"I don't want people panicking," says Wallace. "I want them thinking, 'That loud guy with the white hair says this is normal.'"

He's made his career as a kind of minister of safety. He preaches the importance of an emergency kit—what people in the industry call a "go bag." The state emergency management office's position is that residents there should have enough resources to last two full weeks. They advise people on the Olympic Peninsula to have supplies for a month—that includes not just food and water, but also medications, first aid supplies, toilet paper, and clothing or temporary shelter capable of weathering the elements.

That's because tsunamis don't just wash away homes. They also tear up roads, strip power lines, ruin water supplies, and spew toxic waste.

"The human mind cannot grasp the effects of a 9.0 earthquake," says Wallace. "Not even really if you've been in one."

He sees his job as primarily one of marketing. "We're in the business of selling safety," he says. "We have to make it appealing."

He gets mad when people say the public is apathetic, that they don't listen to warnings. "They will if you package it right," says Wallace. "Sometimes you have to rattle cages."

We get up to leave. As we do, he scans the nearly empty restaurant.

"Who here has bottled water in their car?" he asks loudly. "If you don't, you've already got a problem."

The mayor from a neighboring town, who is having a quiet lunch with his wife, looks at Wallace, sighs, and raises his hand. He's heard Wallace's schtick before. Wallace gives him a thumbs-up as he holds open the door.

~~~

Not enough people talk about earthquake success stories, says Wallace. He wants me to see one of the biggest. So we make our way back to his car and cross a bridge toward the town of Westport. The road we take is on a narrow spit of land just a few feet wider than the asphalt. It wouldn't take much for water to overtop it.

"One way on, one way off," says Wallace. "And look up there." He points to power lines. "One grid. One set of lines."

Just a few more miles, and we reach the Ocosta school complex. It's your typical arrangement: a sprawling high school, a fleet of buses parked under a pavilion, administrative offices and athletic fields. They all sit in a geological bowl. It's a half mile to the beach. During winter storms, water laps at the football field. Once it's in, there's no place for it to get out.

The region's connection to the ocean is everywhere. Nearby, a massive harbor houses halibut and tuna boats. A lot of the kids have parents who work on one of them. Huge black-and-white photos of crabbers and clammers bedeck the school halls.

Paula Akerlund is the school principal. She took the job not long before the Tohoku earthquake. She says she and her staff watched the subsequent tsunami as it rushed over the world's biggest seawall, then shattered windows and doors, flooding homes and schools.

They saw the way those floodwaters lifted entire cinder block buildings off their foundations and swept away buses and cars. The topography around Westport isn't so different. Could the same thing happen there?

Akerlund read FEMA manuals and state statutes. She consulted geological histories.

In 1949, an M 7.1 quake ruptured near Olympia—about seventy miles from Westport. Luckily, schools were on spring break when it struck. Still, it did a fair amount of damage to thirty schools. And two kids died that day. That could still happen today.

Unlike California and Oregon, Washington State doesn't require seismic inspection of school buildings. Nor are schools required to withstand a tsunami. About 30,000 kids live in the inundation zone. A recent FEMA study predicts that an M 9.0 quake would collapse about 75 percent of Washington's schools. Most of the ones left standing aren't high enough to escape floodwaters from a tsunami.

The state estimates it would cost upwards of $13 million just to do a seismic audit of their schools. Who knows how much the retrofits themselves would cost. Most of their principals say they just hope for the best. One of the school buildings in Aberdeen is like that. It was part of a federal study to see how school buildings in the Pacific Northwest would fare in a quake. Inspectors there said it was almost certain the school would see massive damage or even collapse.

That didn't seem like much of a choice to Akerlund.

The district needed a new elementary school. Why not make it a tsunami sheltering space as well? The planned gym was already going to be forty feet tall. Elevate it a little and create the necessary infrastructure, and you'd have a safe place for all the nearly 700 kids in the district.

This isn't as easy as it sounds. About 75 percent of Ocosta students are on the free and reduced lunch program. The county has one of the highest unemployment rates in the state. The tsunami platform would add an additional $2 million to the building bond. Akerlund decided to go for it anyway.

Initially, she says, people were reluctant. The hazard just seemed so tiny and the building was already so expensive. But she campaigned hard. She held open-door meetings with the community and regularly asked the design architect and engineer to make the three-hour drive from Seattle. They did. Repeatedly. Residents began to trust what they had to say. And they listened when Akerlund talked about the safety of the students. These were *their kids*, after all.

On the night of the election Akerlund was nervous. But she'd already decided it was far from the last straw, even if the community defeated the referendum. "I had already told everybody, 'I'm going to run it and run it until it passes. We are not going to have a situation where our kids are not safe.'"

It passed overwhelmingly.

"Want to see what we built?" she asks me. Of course I do. So she and Wallace and I don hard hats and wander over.

The first time I visited, the school was still under construction— waiting for some cosmetic work, mostly, but plenty done to understand just how significant this project is. Four massive stairwells lead up to the tsunami shelter on the top of the gym. Each contains 50,000 pounds of rebar and can accommodate a flood of kids and teachers. The doors to each are connected to the NOAA tsunami warning center. At the first sign of a wave, they open automatically. Surrounding those stairwells are concrete-enclosed steel columns. Those nearby buses are going to become battering rams; the stairwells need to be secure enough to withstand them.

Up top, the district has stocked enough food for a week—maybe more, if they stretch it. But Wallace is hopeful the students would be some of the first people rescued.

"They're kids after all," he says.

Surrounding the platform is a six-foot-high parapet. I had to stand on my tippy-toes just to get a glimpse of sky beyond it. I asked about it, assuming it was there to keep the kids corralled. No, said Akerlund. It's mostly to protect the kids from witnessing the devastation. "We know we'll have to introduce them to that eventually," she says. "We just don't want it to be the first thing they'll see."

In June of 2016, Akerlund and her school board hosted the dedication of the new Ocosta school. The gym was packed for the speeches, but the crowd soon spilled outside to watch a national guard Black Hawk helicopter as it hovered overhead, demonstrating how kids could be plucked from the roof.

Today, two helmets sit on Akerlund's windowsill next to pictures of her corgis. One is an Ocosta football helmet emblazoned with their mascot, the Wildcats. The other is her project hard hat.

NOAA says she and her students would have twenty minutes to make it to the rooftop shelter in the event of a local tsunami. They've been drilling ever since the school opened. Akerlund says, by now, they've gotten it down to eight—maybe a little less on a good day.

While I was there, a parent stopped by to thank Akerlund for the new school. The parent told us she commutes to Olympia every day for work. She'd thought about quitting—it just seemed so far away, should an earthquake strike—but jobs are few and far between in Westport. So she kept hers, but worried like crazy every day. The new school changed that. "My kids are okay now," she said. "I can go to work and know they'll be safe."

PREDICTING THE UNPREDICTABLE

On March 17, 2014, Emily Brodsky and Thorne Lay were having coffee in UC Santa Cruz's Seismology Lab. The two scientists do this pretty much every morning. They call it "seismo coffee" (even though Brodsky usually drinks tea), and they invite bedraggled graduate students to join them in a well-worn conference room cluttered with academic journals, Tupperware containers, and a few beer bottles. Mostly, the two talk about their research or unpack new scholarly articles. On this particular morning, the handful of doctoral students sitting around the table instead listened as Brodsky and Lay speculated about a series of earthquakes off the coast of Chile that had begun the day before. Could they be foreshocks heralding a megaquake?

These quakes were occurring not far from the Tarapacá region of Chile in a part of the Pacific known for having what geologists call a "recognized seismic gap," which is to say a place known to be capable of quakes but one that has not had an event for quite some time. It usually means that pressure and energy have been building up there for some time and are just waiting to explode.

Brodsky says she and Lay had a feeling that release was about to occur—and in epic fashion. "We were asking each other daily, 'Has it happened? Has it happened?'"

And then it did happen.

On April 1—sixteen days after the first tremor—a major quake rocked Chile. Residents in Iquique, Tarapacá's capital city, reported a cacophony of sounds created by the magnitude 8.1 quake—collapsing buildings, causing landslides, tearing apart roads—before they were drowned out by the sound of tsunami warning sirens. By the end of the day, six people were confirmed dead—a number that no doubt would have been higher, had the epicenter of the quake been near a more densely populated region of the country.

As the number of aftershocks continued to grow, Brodsky and Lay began work on a scientific paper about those first tremors and whether they might have been what seismologists call nucleation, or precursory activity before a big quake. If so, then maybe similar events could be used to forecast future seismic activity.

Prediction is the holy grail of the geological world. It's also a four-letter word in that community and the easiest way to start a barroom brawl among seismologists. British geologist John Milne, one of the fathers of the field, wrote in 1880: "Ever since seismology has been studied one of the chief aims of its students has been to discover some means which would enable them to foretell the coming of an earthquake." This was important, he wrote, not just for scientific integrity but also for appeasing the layperson, who, as it turned out, often held the purse strings to scientific inquiry. "What the public imagine they would like to know about an earthquake is the time at which it might occur. If this could be stated, and at the same time something about the character of the expected disturbance in earthquake districts, seismology would be liberally supported."

By Milne's time, science had given it a serious college try: They'd tracked cloud formations and changing weather patterns, lunar tides and the behavior of goats. In the decade or two that followed, they'd also investigate the tilt of the earth, the presence of magnetic fields, the temperature of hot springs, and the possibility that the earth may well have an arrhythmic pulse.

With the Cold War still very much raging, the United States and the Soviet Union entered into a joint earthquake-prediction program in 1972, which was announced at a bilateral summit hosted by Nixon and Brezhnev. Widely touted as the "most successful scientific partnership" between these two chilly nations, the project brought together American and Soviet geologists in the name of wresting some control back from the planet. Why? Because, the scientists wrote in their first official report, every year our ability to increase geological hazards grows larger. They cited mining, the capture of fossil fuels, quarry blasting, and growing reservoirs as examples. Natural earthquakes were bad enough. Man-made earthquakes were making things much worse. "Of all geologic hazards," they warned, "earthquakes can cause the greatest loss of life and the largest property damage." In an attempt to thwart this destructive power, they hung out in one another's labs; they blew up stuff together; they visited one another's dams and oil fields and power plants; they nosed around their seismic arrays. The end result was a 529-page document entitled *The Soviet-American Exchange in Earthquake Prediction*, which was, unfortunately, very light on the subject of earthquake prediction.

Just to be safe, we also launched our own earthquake-prediction program. To the casual observer, it seemed to be going swimmingly. In September 1975, *Time* ran a cover story boasting that leading nations, including the United States, China, and the Soviet Union, were consistently predicting quakes with impressive accuracy. The US National Research Council urged Congress to fund a long-term program aimed at promoting seismic prediction. It was a decade of profound optimism and more than a little scientific hubris—the same decade Nixon announced his war on cancer and predicted a cure would soon be found, the same decade scientists predicted that all the world's oil would be depleted by 2000.

But sustained seismic prediction soon proved more difficult than anyone had thought.

The Soviet Union invested heavily in earthquake prediction beginning in the early 1950s, and their Complex Seismological Expedition was viewed by many as the best science on the subject. In 1969, a group of US geologists traveled there and were so impressed with the Soviets' findings that they snapped photos of all their diagrams and research. Japan had been conducting similar research. The real leader in the field, though, appeared to be China. There, government-funded projects included charting everything from water table levels to electrical currents below the surface of the earth to, famously, the behavior of wild and domestic animals.

Then, in February 1975, the Chinese government issued an earthquake warning and evacuation for the area surrounding the city of Haicheng. Within hours of its issuance, a magnitude 7.3 earthquake struck the region, destroying over 90 percent of the buildings there. Just over 2000 people died. A lot, to be sure, but scholars estimated the death toll would have been 150,000 had the warning not been issued.

Prior to the earthquake, residents reported seeing snakes rise out of the ground and livestock behaving "strangely," though there was never much said about what that behavior looked like. Nevertheless, that was enough for China to launch a study on animals and earthquake prediction. The USGS started to go down this road as well. They even sponsored a symposium on the subject in 1975 (though you'd be hard-pressed to find any evidence of it on their website or in their promotional materials). Without any major quakes, it was hard to say anything definitive. And yet examples of animals anticipating earthquakes continue to make the news. Just five or ten minutes prior to that M 5.6 earthquake that struck Oklahoma in 2016, weather radar captured tens of thousands of birds taking flight. Zookeepers at the

National Zoo reported similarly atypical behavior of their charges, including lemurs, flamingos, and Iris, the charismatic star of the orangutan exhibit, who began screaming and then climbed to the top of her enclosure in a panic just minutes before the 2011 quake struck there.

In general, seismologists remain skeptical about the degree to which any animal can predict an earthquake. And certainly the Chinese prediction program has witnessed some major failures. In 1976, a magnitude 7.6 earthquake rocked the city of Tangshan. Scientists never saw it coming. Neither did the approximately 1 million people living in the city. A quarter of them died that day. Another 164,000 were injured. That same year, the Chinese government issued another warning for a major quake. Residents streamed out of their homes and camped in fields and streets. The earthquake never came.

However, neither of these incidents did much to cool the prediction fervor. The Soviet Union was producing its own excitement in Garm, an ice-swept section of Siberia that had witnessed a major quake in 1949 that killed 12,000 people. Since that disaster, Soviet geologists had focused their study there. In 1971, they announced that they were able to anticipate quakes by examining P wave activity. US geologists began learning Russian, figuring their Cold War nemesis had the best lead on predictive techniques.

In November 1974, seismologist Malcolm Johnston wowed members of California's Pick and Hammer Club when he predicted that a quake would soon hit the small town of Hollister, located east of Monterey Bay. The next day—Thanksgiving—an M 5.2 quake struck the area just as families were carving their turkeys and mashing their potatoes.

Congress approved the US National Earthquake Hazards Reduction Program in 1977. Almost half of its $30 million budget was reserved for prediction research. "People were measuring everything from cockroach activity to radon emissions, seismic patterns to strange electrical currents," John Filson, former chief of the USGS Office of Earthquake Studies, told *Earth* magazine.

Most of the systems, though, relied upon mathematical theories and series of dichotomous keys—statements like "the area has a history of earthquake activity" to which researchers can answer "true" or "false." UCLA's Vladimir Keilis-Borok quickly became the recognized leader in the field. The Russian-born geophysicist founded the International Institute of Earthquake Prediction Theory and Mathematical Geophysics in 1981. There, he and his team used everything from chaos theory to geodynamical statistics to create an algorithm capable of detecting earthquake patterns. To test the program, he also began using it to predict US presidential elections. He died in 2013, but his colleague Allan Lichtman, who helped to develop the election keys, continues the presidential-election-prediction work. They still maintain 100 percent accuracy—including the successful prediction of Donald Trump over a month before his upset victory over Hillary Clinton.

The earthquake side of this work can't boast the same batting average. To date, they've successfully predicted just one earthquake— the M 8.1 quake that struck Hokkaido, Japan.

Allan Lindh has been at the center of US earthquake prediction since its inception. Lindh joined the USGS in 1970 and was a reluctant scientist. He'd spent most of his teens and twenties traveling the West Coast with a tattered copy of Jack Kerouac's *On the Road* as his guide. He eventually landed in British Columbia, where he stacked lumber in what he calls "crummy little sawmills" before meeting a geologist who told him there were easier ways to earn a living. A few months later, he arrived at Stanford. Ten years after that, he had a PhD in geophysics and a position with the USGS.

Lindh, who is now seventy, says he was a "young punk—a nasty, obnoxious person" when he entered the federal agency. This is hard to believe of a man who wears a long white beard and even longer whiter locks in the spirit of Walt Whitman or Charlton Heston's Moses—a man who spends most of his time today tending to his laying hens and writing polemics against nuclear war, whose LinkedIn profile lists his current occupation as an "apprentice at life" and whose telephone number I found, incidentally, on a website dedicated to meditation groups in what may be California's crunchiest town. What is easier to believe is that the USGS was, in Lindh's words, a largely autocratic organization when he was hired. In 1977, he sat in one of the organization's field offices and listened as his supervisors detailed the list of projects they'd be undertaking in the coming year.

"I stood up and told them they should be ashamed of themselves. It was the dumbest list of experiments I'd ever seen. The USGS is supposed to be about public interest, not esoteric science. My supervisors told me if I was so smart I should go write my own project proposal. So I did."

What resulted was the Parkfield prediction experiment.

The town of Parkfield, California, isn't much of a town at first glance: an unincorporated community located in the Central California mountains about halfway between LA and San Francisco, it has a population of either thirty-four or eighteen, depending on whom you ask. There isn't much in the way of infrastructure there either. The nearest gas station is twenty-five miles away. The closest grocery store is even farther.

But what Parkfield lacks in amenities, it more than makes up for in seismic activity. Parkfield sits atop San Andreas. That in itself isn't all that unusual in the Golden State, but the particular patch of fault on which this little ranching community sits is. Parkfield exists at the nexus of two very different types of fault activity: To its south, San Andreas is loaded, much like the Red Canyon and Hebgen Faults were before they erupted in 1959. The section of San Andreas north of Parkfield, on the other hand, is slowly creeping. This nexus of activity offered Lindh and other seismologists something they couldn't find anywhere else: reliable earthquakes. A magnitude 6.0 earthquake had occurred there in 1966. And in 1934. Also in 1922 and 1901. Logic dictated another one would happen around 1988.

That offered a once-in-a-lifetime opportunity: to record and study an earthquake in real time—and, more important, to figure out what happens in the days and hours *before* the quake.

"We figured it was shooting fish in a barrel," says Lindh.

Working with the state of California and several branches of the university system there, Lindh and his colleagues at the USGS installed arrays of instruments, including devices that sound like they belong in Adam West's Batcave: creepmeters, proton magnetometers, motion accelerometers, trilateration arrays. They measured strain and ground deformation. They studied the earth's magnetic field. Basically, says Lindh, they installed any instrument they could think of that might contribute data to the study of how quakes begin and behave. The tiny town began to look like a NASA flea market.

"We told the people at Parkfield we didn't know what we were doing," recalls Lindh. "They kept their cool. They knew we were just doing our job."

More than that, says Lindh, the residents of Parkfield were exactly the kind of people to get behind this sort of thing. "They weren't the kind of people who were going to freak out the minute you started to predict earthquakes around them." Instead, most of them thought it was pretty great. The town had a one-room schoolhouse that looked like a decrepit mobile home. Its lone employee, who also served as a de facto mayor, was charged with manning some of the instruments. Other ranchers offered up their land for further instrumentation. There was a kind of ebullience around the whole thing, says Lindh. It felt like a community effort.

Jack Varian is one of those residents. He and his wife, Zera, moved to the area in 1961, after learning the hard way that mountain ranching is a lot more struggle than it is success. They made a decent go of it until drought struck in the 1980s, making ranching hard even in places like Parkfield. With few prospects and four kids, the couple decided they needed a new plan to breathe life into the region.

For over a decade, the area's only tourism was predicated almost entirely on the untimely death of James Dean: He'd been driving his Porsche convertible, nicknamed "Little Bastard," to a race in Salinas when he T-boned another driver just outside of town. The Hollywood star died en route to the hospital. A memorial quickly sprang up just down the road from the scene of the crash and just across from the Jack Ranch Cafe. People came to pay their respects and eat tuna melts. Varian figured he could start a café of his own and built the Parkfield one in 1985. It wasn't much of a café, really—just a barbecue pit outside a log cabin decorated with junk he and Zera found around the area.

His timing was far from random. That same year, Lindh and the Parkfield project issued its official prognostication: Parkfield would see a magnitude 6.0 quake by 1993. Given the historical regularity of quakes there, this was far from an audacious claim. But it was revolutionary if for no other reason than that US scientists had never made a prediction like this before.

The prediction, says Lindh, was based on a single computational model. He and the rest of the Parkfield team took it as a given that the next Parkfield earthquake would look like the others—that it would rupture in the same direction, and that it would be preceded by the same sequence of foreshocks. They assumed that sections of fault like the one at Parkfield have an identifiable threshold for stress—that you can depend upon it snapping whenever that threshold is met. There was no reason to believe that wouldn't occur in the same time frame it had for over a hundred years there.

Varian liked the sound of it. He came up with the town's motto, "Be here when it happens," which he painted in large letters on the iconic water tank outside the café. He built a seven-room guest lodge with the slogan "Sleep here when it happens." And he orchestrated the café's menu accordingly (these days, its most popular item is the "Shakin' Burger," made with beef raised on Varian's ranch).

Years went by. People continued to wait for the Parkfield quake. Meanwhile, the state of California's Office of Emergency Services began dabbling in short-term earthquake warnings. They issued their first one in 1985 when the USGS saw an increased risk of a moderate quake occurring in San Diego. The state issued another advisory in June 1988, after a magnitude 5.1 earthquake erupted near San Francisco. The USGS announced there was a 5 percent chance another would follow in the next five days. It didn't. They issued a similar five-day prediction after a similar quake rumbled through the region in August 1989. The quake didn't come in the window promised, but when Loma Prieta slammed the region two months later, people began wondering if there was something to the process.

Meanwhile, a New Mexico–based climatologist named Iben Browning claimed to have predicted the Loma Prieta quake using computations based on the earth's gravity and tidal patterns. Halfway across the country, the USGS had become simultaneously interested in the New Madrid fault. The organization published estimates stating there was anywhere between a 13 and 65 percent chance the fault would rupture again before 2000. Browning took it one step further: He predicted there'd be a major quake there on December 3, 1990. Ads began appearing in papers like the *St. Louis Post-Dispatch* proclaiming Browning a "remarkable man" and including a quote from Southeast Missouri State University geologist David Stewart calling Browning "the most intelligent person" Stewart had ever met. For just $39, readers could purchase a VHS tape with thirty minutes of excerpted interviews with Browning in which he explained his predictive process. The full ninety minutes could be purchased for a mere $99. I called the 1-800 number listed on the ad and got a vague recording asking me to call back—no matter when I called.

Browning's theory was based on ideas of tidal loading, which is to say the way in which tides redistribute ocean mass and the subsequent shift in stress that occurs. Browning had noted that, on December 2 and 3 of 1990, high tide would occur when the earth and moon were at their closest point and when the earth, sun, and moon were in alignment. These conditions had occurred a few other times that century without any noticeable increase in seismic activity, but that didn't dissuade Browning. Neither did the resounding rejection of his theory sent up by the seismological community. Geologists and seismologists at regional universities all made public pronouncements against Browning's claims. An ad hoc community comprising scientists from the USGS, the California Division of Mines and Geology, and the Argonne National Lab did the same. They all cited the lack of precedent for this theory—and Browning's clear lack of scientific method.

Nevertheless, media coverage of his prediction exploded. Days before the predicted earthquake, dozens of news trucks, tilting under the weight of their giant satellites, parked in every available space in the tiny town of New Madrid. Reporters from tiny local papers rubbed elbows with correspondents from *The New York Times*, the *Miami Herald*, the *LA Times*, *Newsweek*, and *Time*. Over 200 media outlets in total had camped out in the town, including *Good Morning America* and the *Today* show, both of which scored interviews with Browning. Even the BBC showed up.

Local emergency managers took note. I remember. I was a sophomore in high school then, and the usual tornado and fire drills were suddenly augmented by earthquake protocols as well. We did a lot of those drills that year. Communities around the region, including the cities of St. Louis and Louisville, participated in their own disaster drills. State national guards called up their troops. Homeowners rushed to buy earthquake insurance. Then, on September 26 of that year, the town of Cape Girardeau, Missouri, experienced an M 4.7 quake: Clearly, speculated many in the region, this was a foreshock heralding the validity of Browning's prediction. The frenzy intensified.

By December 2 of that year, the town of New Madrid—unsure whether to plan for a party or a mass funeral—had decided to embrace both. The chamber of commerce sold T-shirts and coffee mugs; local restaurants trotted out earthquake specials; a bluegrass band with the name Faultline Express played on a stage erected expressly for them. Meanwhile, other residents hung "SOS" banners from their homes and sat for doomsday sermons given by local preachers. Rumors that the state had ordered truckloads of body bags prompted thousands of parents to keep their kids home from school.

December 3 came and went. The local McDonald's handed out free coffee. The Red Cross distributed earthquake preparedness kits. The quake never manifested. David Stewart resigned his post as director of the SMSU earthquake center. Iben Browning began to fade into oblivion. He died of a heart attack six months after the quake that never happened.

⁓

The same year Browning made his prediction, Alan Lindh was more confident than ever about Parkfield, California's chances for a big event.

In 1990, Lindh authored an article in *MIT Technology Review* in which he put the chance of an earthquake hitting Parkfield by 1993 at 95 percent. He was interviewed by *The New York Times* in January of that year. The reporter characterized him as a "modern-day Jeremiah," one of the major prophets of the Old Testament. Lindh corrected him: "It's not like being Jeremiah," he told the reporter. "It's like being a high school teacher explaining AIDS to teenagers." *Times* readers took the comment in stride.

Months went by. Still no earthquake.

Andrew Michael was a field geologist for the USGS working in Parkfield then. He says the overall footprint of the project was never all that big: It began with two technicians working in a small ranch house. As the years went on, they were down to one employee and a single trailer home. Most of the data from the instruments was, instead, patched back to the USGS offices in Menlo Park. Geologists like Michael wore pagers that would alert them to any activity.

When a Japanese film crew arrived in Parkfield to document the experiment, they were incredulous—and more than just a little suspicious.

"They kept demanding, 'Why are you hiding your control room from us?'" says Michael. "It was inconceivable to them that we didn't have one."

Meanwhile, project supervisors were under intense pressure to step down the experiment even further.

"A lot of people wanted the money for Parkfield redirected for other things," says Michael. "We had one of the best strong motion arrays there, and there were definite pressures to move it to Southern California where there were a lot more people."

The National Earthquake Prediction Evaluation Council responded by conducting an extensive review of the Parkfield project. They concluded that, even though Parkfield hadn't yet seen another quake, it was still "the most likely place yet identified to trap a moderate earthquake in a densely instrumented region and the best locale identified to answer a number of important scientific questions about the seismic source." The project, they recommended, should be continued. A 1995 fact sheet issued by the USGS promised that earthquake forecasting had advanced from a research frontier to an emerging science—and one "reliable enough to make official earthquake warnings possible."

For his part, Lindh began reciting his favorite quotation by Charles Richter: "Only fools, liars, and charlatans predict earthquakes." It's still a line he uses a lot today. Michael says he gradually began focusing on other projects. But as far as he was concerned, Parkfield was still the mostly likely place for a next earthquake.

They finally got it on September 28, 2004. Michael remembers feeling surprised.

"I was like, '*Really?*' The earthquake wasn't something I was thinking about very often at that point."

Lindh was. He says that, even though the quake was over ten years late, it happened. And it brought with it the opportunity to do some real on-the-ground science. Seismologists learned a lot from the delayed quake, in terms of what it looked like both as its own event and as part of a historical sequence. The 2004 quake ruptured differently than its predecessors, in terms of both the geographic direction of the quake and the rumblings preceding it. Those variables, says Lindh, have offered important contributions to seismic modeling and assessing hazard. If we are ever going to predict quakes, he says, we're going to need a lot more in the way of sophisticated modeling to do so.

IS THE SKY FALLING?

In 2008, the Italian town of L'Aquila, the capital of Italy's Abruzzo region, began to experience an earthquake swarm. At first, most of the tremors were tiny, but as the months continued, they began to grow in intensity. So, too, did the anxiety of L'Aquila residents. That anxiety reached a feverish pitch when Giampaolo Giuliani, a technician at the Gran Sasso Nuclear Physics Laboratory, began issuing his own earthquake warnings.

Giuliani had long been interested in the connection between radon and seismic activity. In the late 1990s, he read about Russian scientists who were attempting to draw a correlation between increased radon levels and earthquakes. At the time, he was working as a lab technician for the National Institute of Nuclear Physics, where he tended instruments dedicated to cosmic rays. But what Giuliani really wanted to do was study the energy inside our own planet. He requested a transfer to the national geophysics lab and formal funding to study radon and seismicity. That request was denied, citing a lack of scientific integrity. So Giuliani began building his own radometers, which he distributed around L'Aquila. The mayor of the town agreed to let him put one in the school's basement. By late March 2009, Giuliani had noticed a significant spike in radon levels. He called the town mayor. He also published his findings on the web. That didn't go over well. He was served an injunction and ordered to take down the warnings. But Giuliani kept at it privately.

When a professor at the Open University noticed that a huge colony of toads had disappeared from a nearby lake, this seemed to strengthen Giuliani's case: It was as if the toads had noticed the increase in gas as well and decided to hit the road. Similar phenomena had been observed prior to several earthquakes in China: The amphibians and reptiles either appeared or disappeared en masse before a quake. The phenomenon even prompted a German chemistry professor to publish a book entitled *When the Snakes Awake*, which argues that tiny changes in particulates before a quake prompt a noticeable shift in animal behavior. The book was published by MIT Press, but was received very much as earlier animal prediction theories had been.

That's pretty much happened with Giuliani as well. In the late 1970s, a team of researchers at Caltech had attempted to prove a similar correlation, but they eventually concluded it was impossible: True, radon levels increased during some quakes, but they also remained the same during others. Susan Hough is one of earthquake prediction's biggest skeptics. "Radon levels," she writes in *Predicting the Unpredictable: The Tumultuous Science of Earthquake Prediction,* "have a knack for bouncing up and down for reasons other than impending earthquakes."

Nevertheless, Giuliani remained confident in his findings. And his claims had gotten the attention of neighboring towns. One mayor even went so far as to dispatch vans carrying loudspeakers to warn residents.

That did not impress Italy's National Committee for Forecasting and Preventing Great Risks. Nor did it delight Guido Bertolaso, director of Italy's Civil Protection department. Bertolaso was understandably tetchy about Giuliani's claims and the traction they were getting. It looked like things were unraveling.

Residents of the region were confused. And confusion makes for even more anxiety. Bertolaso summoned seven experts—three seismologists, a volcanologist, two earthquake engineers, and a hydraulic engineer—along with local officials and members of the great risks committee. He also made the mistake of telling a local official that the meeting was a media operation to "shut up" Giuliani (but that didn't get out until much later). The official aim of the meeting, as stated in the letter sent to the scientists, was to undertake "a careful analysis of the scientific and civil protection aspects of the sequence ongoing for four months."

Later, the scientists in the room reported that they had explained there was no existing way to predict earthquakes—that swarms like the one L'Aquila was experiencing are common. Sometimes they lead to bigger quakes; often they don't. That's all true, says American seismologist Tom Jordan.

But in the press conference after that meeting, Bernardo De Bernardinis, the deputy head of Italy's Civil Protection department, interpreted the scientists' conclusions differently: "The scientific community tells us there is no danger, because there is an ongoing discharge of energy. The situation looks favorable."

Prior to that all-important meeting De Bernardinis gave an interview with a local journalist. He told the journalist that there was no reason to raise further alarm. So should we all have a glass of wine? asked the reporter. Sure, said De Bernardinis. That interview was broadcast after the meeting. It led many in the region to assume it came out of the meeting—that De Bernardinis made these claims after hearing what the experts had to say. (It also became known as the "glass of wine" interview, which wouldn't do a lot to help De Bernardinis's credibility later).

A week later, in the early-morning hours, disaster struck. A shallow M 6.3 earthquake rocked the region. The effects were severe. Over 300 people were killed. Some 80,000 were left homeless. Survivors wanted answers. More than that, they wanted retribution. Families, they said, had gone to bed thinking they were safe. They would have taken precautionary steps if they'd known otherwise. (Giuliani and his family, incidentally, went to sleep fully clothed that night. They also opened all their windows and doors so they could leap out of the building if necessary. As it turns out, it was.)

Immediately after the quake, six of the scientists were put under official investigation for offering false assurances to the Italian people. So, too, was De Bernardinis. That investigation resulted in manslaughter charges. What followed was a five-year court battle that many scientists around the world saw as nothing short of absurd. Prosecutors contended that the scientists' assessment of earthquake risk to L'Aquila "was approximate, generic, and ineffective with reference to activity and duties of forecast and prevention." More importantly, they alleged that these scientists communicated "incomplete, imprecise, and contradictory information on the nature, the causes, the hazard and future evolution of the scrutinized seismic activity" to the National Civil Protection department and to the people of L'Aquila.

All seven individuals were found guilty and given six-year jail sentences, along with fines of 1 million euros each and a lifetime ban on holding any position in the public sector. They appealed. The scientists were acquitted, in large part because they didn't actually communicate beyond the meeting.

De Bernardinis was not. Prosecutors argued it was his job, not the scientists', to apprise the public of any hazard—or lack of hazard. In their view, his statements to the journalist wrongly reassured the public. The judge agreed but did reduce his sentence to two years.

The problem, though, is that scientists don't know whether these kinds of little quakes are foreshocks until after a big quake occurs. Plenty of minor events dissipate without a major event. Some main shocks occur with little or no observable precursory activity. As Jordan explains it in the case of L'Aquila, "A big shock was not very likely: the probability of a false alarm (if an alarm was raised) exceeded the probability of a failure-to-predict (if an alarm was not cast) by a factor of more than 100."

This was the same problem Lay and Brodsky pushed up against in their paper about the 2014 Chile earthquake, which they eventually published in the journal *Science*. There, they concluded that while the foreshocks for the Chile quake may suggest that there are "observable precursors" to an earthquake, they lacked the kind of measurable distinction needed to be clearly identified as anything other than just random seismic activity. To establish that kind of pattern, says Brodsky, seismologists need a lot more data—and the ability to analyze it for patterns. Most large earthquakes occur far belowground—and often in areas that are thousands of feet under water, making it all but impossible to locate data-collecting devices close enough to record the small seismic activity preceding a major quake.

Without that kind of data, says Brodsky, earthquakes will remain the least understood natural disaster on our planet.

Brodsky is an upbeat, encouraging human. Her students adore her. She's funny with a kind of New York wit that borders on sarcasm, which seems exotic and novel in a place like Santa Cruz. You can bring her just about any theory in the world and she'll consider it. When I first found her, she was giving a talk at the National Science Foundation, trying to explain what is known about why earthquakes happen. She brought props: a brick and a piece of wood and a bungee cord and a plastic giraffe. I don't know exactly what the giraffe was supposed to symbolize (maybe the desperation survivors feel when they realize FedEx can't evacuate them?), but as I watched him topple over at what was supposed to be a very fancy luncheon for very important people, I knew I had my seismologist (though I did later learn that Thorne Lay puts a picture of a kitten in every one of his PowerPoint slides, and that's pretty endearing, too).

As it turns out, Brodsky isn't technically a seismologist. She's actually a geophysicist, and a very smart one at that (even by geophysicist standards). Her areas of expertise include the processes that trigger faults (full disclosure: she's the one egging on Heather Savage to trigger a bigger earthquake). And in the world of people who study this kind of thing, she's one of the most amenable to the idea that earthquake prediction is even possible.

"Earthquakes happen from a known stress," she told me. "So the question is, how does that translate? Can we translate it into something a little bit more practical? Like a prediction?"

For her, the Tohoku earthquake was a watershed moment. It was so enormous—so unexpected. Perhaps more importantly, it happened in a place with lots of sensors and lots of people paying attention. But it's still not enough to advance prediction science.

"Maybe," she says. "Maybe someday." Maybe after we have more and more quakes and the technology to record what happens in the days and weeks beforehand. Maybe if we decide it's worth funding projects like some of the ones I've mentioned in this book.

Maybe data points will begin to emerge. She's written papers about places like the Salton Sea geothermal plant. It does seem possible to predict a quake there, or at least it might be if you know the injection and extraction rates at the power companies. Maybe if we figure out what happens in those creeping events, like the ones at Parkfield. That requires more instrumentation. Part of why she and Lay were able to watch the 2014 Chile quake unfold was because of the amount of instrumentation located in and around the subduction zone. Which of those foreshocks was the precursory event? Which one said a bigger earthquake was going to follow? That, says Brodsky, is what we really need to know.

Her colleagues tease her about this sometimes.

Just weeks after the L'Aquila quake, the Italian government convened an international panel of seismologists to discuss how they can improve earthquake prediction in their country. They asked Tom Jordan to chair. "Our report reaffirmed that there is no known method to predict earthquakes with high probability," he wrote in *New Scientist*.

Besides, he says, it's missing the point. Three prominent Italian seismologists agree. Instead, they argue, scientists should "insist on passing the message that in earthquake prone areas there is a permanent high seismic risk, that is, a 'permanent emergency' motivating the need of communicating risk and emergency management to society." Only that kind of communication, they conclude, will force the kind of "prevention and preparedness initiatives" capable of mitigating any seismic risk.

Brodsky agrees. And she says it'll take years, maybe even longer, before we even know if it's possible. But she's also going to keep looking. Maybe, eventually, we'll find something.

YOUR TWO-MINUTE WARNING

Inside the Berkeley Seismology Laboratory conference room, Jenn Strauss is demonstrating how a beta early warning application works. It's a warm afternoon, and she's wearing a denim blouse knotted over a cotton skirt and flip-flops. She's really excited and hops around as she talks about this new early warning application. As the minutes tick down, we both watch her cell phone, waiting for the imaginary text alerting her to an earthquake. "Here we go," she finally says. Her cell phone becomes a stopwatch, counting down seconds until our imaginary quake. She calls them out, counting the seconds, as she hops over to the bookcase and grabs the biggest book she can find. It's supposed to represent her young daughter. Then she walks just a little more briskly to a dry-erase board and grabs a marker. "Go bag," she explains. Then she crawls, still carrying the imaginary daughter and bag, under the large boardroom table. She hunkers down there for just a second or two before peeking her head out. "We made it," she says with a grin.

Strauss apologizes—this demonstration is unusually low-tech, especially for Berkeley. The application on her phone is anything but. Instead, it's part of ShakeAlert, our nation's nascent early warning system. In 2006, a consortium of state and federal agencies, along with universities in California, Oregon, and Washington, began work on this project. Currently, it taps into information generated by about 700 seismic stations along the West Coast to provide early warning of an impending quake.

Some people would say this technology is long overdue. Mexico has had an early warning system for years; that country began work on theirs immediately after the devastating 1985 quake. Two years after it was fully operational, an M 8.0 quake struck off the coast. The early warning alert system began to issue public warnings. Schools began to evacuate. The train system shut down. People survived. Were a similar one to strike again today, residents of Mexico City would receive as much as ninety seconds to prepare.

Beginning in the 1960s, Japan began a similar system to alert its bullet train operators to an increase in seismic activity. They dramatically expanded this system after the 1995 Kobe quake. Today, alerts are broadcast over radio and television and in high-traffic areas like stadiums and movie theaters. Pop-up windows appear on laptops and devices. Individuals can subscribe to text updates similar to the one Strauss is showing me. Additional systems are in place in countries like China, Taiwan, Turkey, and the Philippines.

The basic premise underlying all of them is fairly simple: Radio waves travel faster than ground waves. Recall that an earthquake is heralded by two waves: the P wave and the S wave. The P wave, you may remember, travels faster than the S wave. It'll arrive at a field seismometer first. Once it does, that seismometer relays information about that wave to a central computer. It, in turn, can issue a warning that is pushed out to cell phones and other devices before the ground shaking begins. The closer you are to the epicenter, the less time you'll have. But even ten or fifteen seconds is enough to find cover, says Strauss. Her demonstration—with her stand-in daughter and the go bag—took less than thirty.

That would have been enough time for people in the Bay Area to hunker down in the case of the Loma Prieta quake. Strauss says that a warning system like ShakeAlert would have gotten the message out in about five seconds: plenty of time to slow trains, prevent planes from trying to land, and stop traffic in dangerous places. Some schoolkids would have had twenty seconds to get under their desks. People at home and at work could have reached a study table or other safe spaces.

California's second-biggest earthquake occurred in 1857, and some scientists think it's the best model for what might happen in Los Angeles in the future. If that proves true, then residents of that city might even have close to ninety seconds to prepare. And this isn't limited to the West Coast. New Madrid, Virginia, New York, Hawaii, and even Oklahoma would see benefits from this kind of technology. That is, so long as they are some distance from the epicenter (in the case of New Madrid, for instance, folks in St. Louis would probably get more time than those in Memphis). One of the limitations of early warning systems will always be that they need time to get the message out. That would have been in short supply in the 1906 San Francisco quake, where the epicenter was right near the city. By the time people there received word, it would have been too late.

But what, exactly, would people in the impact zone see?

Strauss shows me an example on her phone. It's perfectly simple and clear: a magnitude number, say M 6.0. And then a countdown clock, along with the last-known location in relationship to you. That's important—this only works if the app knows where you are.

The real trick, she says, is calibrating the number of alerts people receive and how they respond to them.

Time is absolutely of the essence for an early warning system to work. A seismometer can send information to a mainframe computer in about a second. But that computer has to run a series of algorithms to verify what it has received. Is it actually a P wave? And if so, how big of a quake is it heralding? Where will the shaking take place? How intense will it be?

The more data points that computer has, the more likely it will avoid either false alarms or missed quakes. But those data points take valuable time to accrue. It takes a couple of seconds to receive data from other seismometers and then again to issue the warning. The more data gathering and number crunching you do, the more accurate your warning will be, but it's also true that that warning will be issued later, and so it may be of less use to the people who need it most.

"It's a trade-off," says Strauss. "We don't want to fall prey to the 'cry wolf' syndrome. If you want to call every single quake, you're going to have some there that are false."

That's what happened in Japan after Tohoku. There was so much energy around, so many signals, that the early warning systems kept sending out notifications about impending quakes that didn't actually happen. For a group of people already intensely psychologically rattled, that didn't do any good at all.

Strauss says that false warnings might not be such a bad thing in places like schools and offices buildings. Even if they don't actually herald an earthquake, they'll provide much needed practice. And practice, she says, does wonders for cutting down on panic. At worst, the schools will miss a few minutes of a lesson; companies will find their meetings momentarily interrupted. It's like fire drills: We may find them inconvenient, but few would argue they are a bad idea.

That's a little more complicated where facilities like airports, oil refineries, and nuclear power plants are concerned. They're expensive to shut down, and doing so comes with a bigger cost to communities.

Right now, the program is set up to send out a warning for any quake with a magnitude of 2.5 or above. That's important, too, says Strauss.

"If we only sent warnings for M 4," says Strauss, "we'd maybe only be sending warnings once every six months or so. That's not enough to test the system."

Maybe more importantly, it's not enough for users to get used to the system.

A lot of it, she says, comes down to social science and what people need. Too many warnings, she says, and people will become too habituated to them—they won't act. Too few, and they'll be surprised by the warnings and may not know what to do.

So far, the technology seems to work. Strauss received a notification in advance of the M 6.0 Napa quake in 2014. It was 3:30 in the morning; she and her husband were both in bed sound asleep. That scenario she showed me with the book and the marker—it's exactly what happened. Except in this case, it really was her daughter and their go bag. The quake did a fair amount of damage to Napa architecture; Strauss and her family weathered the shaking under their dining room table without injury.

Right now, only about ten people have this technology. They're mostly employees of the lab. Meanwhile, hundreds of people are using the ShakeAlert application, including Pacific Gas. BART also has it, and they can use it to slow down or even stop their trains in the event of a quake.

Strauss says the goal is that the general public will also have access to an application. But first, she says, that's going to require education and training. It's one thing to get a notification that an earthquake is coming. It's another thing altogether to know what to do in that instance. A lot of that, she says, is cultural.

"We can't just overlay Mexico or Japan's system here and expect it to work." Those systems, she says, come with certain social behaviors and expectations. People there have trained what to do when they receive an alert; there are protocols for factories and schools and highways. Without them, people tend to freak out. A multicar pileup on the Bay Bridge isn't going to help anyone. Neither is a riot in the New York subway or a mass exodus from the Pentagon.

Recent history has shown in vivid relief what can happen when people receive warnings they don't know what do with. Take the case of Peru. In the late 1970s, at the height of the Parkfield prediction program, a seismologist by the name of Brian Brady theorized that Peru was due for a major quake. Brady had cut his teeth studying rock bursts in mines and had observed definite precursory patterns to these seismic events. In October of 1974, an M 8.1 quake ruptured near Lima. It did a fair amount of damage and took the lives of about eighty people there. Brady hypothesized that a bigger one was yet to come—maybe even as strong as an M 9.2, a quake that would rival the world's strongest and would undoubtedly prove to be one of the deadliest. Brady had successfully anticipated an earthquake near the Salton Sea two years prior. And the explanation for why Peru was due for a megathrust was based on mathematical calculations that some geologists and seismologists found sound. Others disagreed, and a debate among US scientists ensued. News of that debate and, more importantly, Brady's projection worked its way down to Peru. The media there began publicizing it widely. The government got involved. In February of 1980, a representative from the Red Cross in Peru came to the United States asking for aid. Among other items, he asked for 100,000 body bags. That didn't go over very well back home. People began fleeing the country. Home sales plummeted. So did tourism. As the date of the predicted earthquake approached, a kind of hysteria began to take effect. That day, even major cities seemed like ghost towns. By nightfall, there was no earthquake. Nor was there the next day. Or the one after. It took a long time for the Peruvian economy to recover.

That's maybe an extreme case, but it's also a valuable allegory about human nature. When faced with the prospect of disaster, we need a sustainable plan. In the case of an early warning system, that means knowing what to do and when to do it. If I'm working on the twenty-third floor of an office building in Seattle, will I know what to do? What if I'm babysitting a toddler in a brick two-story? Or driving on an overpass?

There's also the matter of how these warnings will be communicated. Ideally, they'd go out on a variety of different channels, including public alert systems, social media, and, of course, individual cell phones. The ShakeAlert consortium has primarily been concerned with the algorithms and data collection needed to issue these early warnings. The nuts and bolts of how those warnings get generated and deseminated remains to be seen. That will require a lot of smart thinking on the part of private firms dedicated to that kind of technology. There was a real groundswell of that kind of business several years ago when the ShakeOut initiative was first launched. Start-ups from LA to Silicon Valley promised they'd be at the forefront of the warning system. Most of them don't exist anymore. The ones that do have diversified to include a lot of other projects.

~~~

Richard Allen has been at the center of America's earthquake early warning project since the beginning. He didn't set out to be an earthquake early warning guy. He graduated from Cambridge in his native England with an interest in modeling the structure of the earth. That's still one of his specialties, but he gets significantly fewer (which is to say pretty much zero) calls from the media about it.

He's one of the people who likes to tease Emily Brodsky, but he does it in a really lovable way. And he admits that early warning systems are a kind of prediction—the kind that's really useful.

"So many people want a silver bullet," says Allen. "They want to know that there's going to be a magnitude 7 quake there next week. Okay. But then what? They can leave, but the infrastructure is still there. If we haven't shored it up, it's still going to be decimated. How are you really benefitting in that case?"

He also points to other natural disasters, like hurricanes, where people do get predictions. Even ones with high probabilities—like Superstorm Sandy hitting New York—often aren't enough to get people to leave, or even to fill their bathtubs so they'll have water.

To get the kind of really, really high probability that might compel people to act, we're going to need a lot more sensors out there. Currently, the ShakeAlert app is relying upon about 700 seismic stations up and down the West Coast. It's not enough, he says, and there are big gaps along the way. Not to mention plenty of earthquakes outside that area. In Japan, says Allen, early warning instrumentation is installed at least every fifteen miles and more like every five miles in areas with dense populations. We'd need that kind of coverage to really have reliable warnings, along with some intensive arrays set up within three miles of every known active—or reasonably potentially active—fault. Allen estimates it would cost about $38.3 million for the initial start-up, along with about $16 million a year after that for maintenance and operations. Even then, blind spots would remain. And seismograms aren't any less finicky than they were when Bob Smith installed them throughout Yellowstone. At best, a single device would probably last about ten years with regular maintenance—assuming, of course, that it stayed up (Mexico's have an obnoxious habit of wandering away, especially in places like the capital).

Historically, Congress has been reluctant to fund early warning programs like MyShake, an earthquake-sensing app Allen has helped develop. There's still a perception that earthquakes are really only a California problem.

Various initiatives have emerged over the years to increase our national seismic coverage in an affordable way. One of the big ones emerged in the 1990s, when Tom Heaton and other seismologists interested in early warning devices debuted a project allowed to gather information from a home, office, or school. Individuals could send away for a small seismic array that plugged into their desktop and then transmitted information back to the consortium. It worked, but never really took off.

Recently, a team of USGS scientists have begun studying ways in which social media might prove helpful. They found that a scan for keywords like "earthquake" on Twitter can often alert geologists to trembling before any official confirmation has been issued—and sometimes within just seconds of the first shaking, far less time than it takes for seismometers halfway around the world to receive the same stimulus. Consider the Tohoku earthquake. It took the USGS National Earthquake Information Center nearly four minutes to receive information about the quake. And even then, they had a loose idea about where it occurred but not how big it was. In just a fraction of that amount of time, tweets from around Japan were providing valuable information about not just where the quake was, but also what they were experiencing. (That's part of what is so effective about the Modified Mercalli Intensity Scale: Something like "all my dishes broke" really means something to scientists—and to the people who are going to have a hard time eating dinner that night.)

Trawling for tweets is not without its limitations, of course. One day the geologists thought there might be a giant swarm of earthquakes around the country, until they realized that Dairy Queen had just announced a buy-one-get-one-free deal on their Oreo Brownie Earthquake Blizzard (no word on whether the subsequent stampede actually created seismic activity, but who knows?). And, they say, scientists can't really know how big the shaking was, or where people tweeting were in relationship to the epicenter. Sometimes, people retweet other people's information, so timing can be an issue as well.

Still, the idea of crowd sourcing got Allen thinking. Or more to the point, it got Allen's graduate student Qingkai Kong thinking. He'd heard about how forecasters had used Twitter to track hurricanes like Sandy and how the Centers for Disease Control was using social media to keep track of how outbreaks of contagious diseases like the flu moved around the world. Couldn't seismologists keep track of earthquakes through crowd sourcing as well?

This was no theoretical question for Kong. He grew up in Central China, in a community right near one of that region's most active faults. He still remembers how his parents and other adults in the town would try to monitor any seismic information they could—how everyone would sleep outside whenever a threat emerged. He did his undergraduate work in civil engineering with the goal of trying to make buildings safer for communities so that people wouldn't have to sleep outside. But as he prepared to graduate, he realized that he really only understood part of the story.

"I had learned about the response of buildings in earthquakes, but I didn't really understand the earthquakes themselves," says Kong. "I wanted to build a bridge between those two sciences."

The earthquake early warning program at Berkeley was the perfect way to do that. He joined Allen's team and became interested in the problem of available seismometers right away—particularly in developing nations.

"Think about it," says Allen today. "An earthquake-prone nation like Nepal has about three seismometers, plus or minus five." (Nepal actually has seven accelerometer stations and twenty-one short-period stations. The Berkeley lab and the regional USGS office together have nearly ten times that amount for an area about twice the size of Nepal. But you get the idea.)

What Nepal does have is millions and millions of smartphones. There are over 600,000 in Katmandu alone. California boasts over 16 million phones. By 2020, there will be over 6 billion smartphones worldwide. Kong wanted to know if they could become the world's densest seismic array. Turns out they can, and without any additional technology. Each smartphone has an accelerometer. It's a sophisticated device that tells the phone how you are orienting it and whether you ought to be seeing something in a portrait or landscape layout. That same tiny piece of technology can also tell people like Allen and Kong whether that phone is experiencing an earthquake.

Kong had to do a lot of tests to figure out whether phones could really pull this off. Turns out they can with remarkable success. They can record all kinds of information about ground motion and the intensity of shaking, too. Working with a team of computer scientists and other seismologists, he also determined that the phones can distinguish between normal human activity (like jumping up and down when your team scores or dancing your fool head off at a concert) and seismic motion. More exactly, the algorithmic programs that Kong helped design can make that determination. Together the team launched an app called MyShake. All users have to do is download it. The phone and MyShake's computers take care of the rest.

Prior to launching the app, Kong and other members of the team sat around that same conference room where Strauss demonstrated the early warning system to me. They were all taking guesses as to how many people might download the technology. One team member figured they'd be lucky if they got a hundred volunteers. Kong's an optimist. He said he thought maybe 200—300 if they were lucky. Another team member guessed 1000. They all laughed at him. Within a week of its launch, MyShake had over 100,000 participants. A year later, that number had tripled. When I visited, Kong showed me a map identifying where active users were at that exact moment (these are folks who have their phones on and the app running). Over a thousand were in New York City. Two hundred were in Iran. Dozens were in Mongolia. *Mongolia.* They're also in the Democratic Republic of the Congo and Venezuela and

Norway and Papua New Guinea. They're in places like Haiti that have a lot of earthquakes but no early warning system. They're also in locations that haven't seen an earthquake in a long, long time. And that's totally fine with Kong: Their data helps him figure out what motion is actually seismic.

I told him I'm surprised so many people are willing to sign up: An app that knows where you are just seems so Big Brotherly. What ensued is an incredibly awkward conversation in which an American journalist attempts to explain that expression to a Chinese seismologist. He was really polite about it. And, he says, the team came up with solutions long before they launched the app. First, they assign a user ID to every app downloader, which anonymizes their information. Data is only collected when there's a triggering event and it's sent directly (and exclusively) to the algorithmic computer. So, in other words, it's impossible to track someone using it. Kong knows he's getting data from a particular place, but he doesn't know who owns the phone. When it comes to maps like the one he shows me, they also add in what he calls "location noise," which is to say they deliberately randomize the data just enough so that an individual cannot be tracked over time.

In time, he says, they'll be able to use the information sent from these phones to issue their own early warnings. And as users increase, they'll be able to create seismic stations not just horizontally, on the surface of the planet, but also vertically. If there were a user on every floor of a seventy-story building, say, they'd get real-time data about how that building is responding in a quake and how the waves are passing through the building.

And here's where things get really wild: Another student of Allen's is studying whether it's possible to create building material that changes to anticipate those waves. Is it possible, in other words, for a *building* to respond to a warning about an impending quake?

So far, the response from app users has been great. They say that they like knowing they might be helping or that their phones might eventually save someone's life. They also are glad for the opportunity to know more about how earthquakes work, since so few of us actually do.

Kong says that's his favorite part about this study. He's about to finish his doctorate. After that, he'd like to do a postdoc. Then he's hoping for an academic appointment. He likes that he can do cutting-edge research there. More than that, he likes the way knowledge tends to spread in those types of places. "You are learning new things every day not just from the class but the students as well. In earthquake science, that is a very, very good thing."

AFTERWORD

Picture this: It's 2:00 A.M., and you are fast asleep. Or maybe it's 2:00 P.M. You're at work; your son or daughter is about to board the school bus and head home after a day of classes. In either scenario, you are probably thinking about anything other than an earthquake. But then the tremors begin, first with a sudden jolt strong enough to knock someone from their chair or bed. What follows is a minute of hard, bone-rattling vibrations—the kind that make even walking or crawling hard.

This quake is a big one—maybe not the *big one*, but it doesn't have to be. It's going to do real damage to the entire region in which you live. The question is this: Just how much damage? And will you and the people you love survive it?

That depends.

Say that we as a nation continue our current course when it comes to infrastructure. Our roads and bridges continue to languish. We invest in initiatives other than levees and the grid and safe buildings. As individuals, we continue to believe that the prospect of a major quake where we live is too implausible to even consider. In that case, the results of this quake are going to be devastating. And after reading this book, you can probably anticipate with some degree of specificity just what's going to happen. Do you live in a building with unreinforced masonry or nonductile concrete? It could come tumbling down. Even if it remains standing, that building could be rendered uninhabitable for months or even permanently. Now you're homeless. Where are you going to go? Do you have the resources to make that happen? Say you're one of the lucky ones: Your house is symmetrical and made of wood; your town house has been retrofitted. The building may well survive the quake. But what's going to happen inside? That bookcase next to your bed or the shelf above it: Are they and their contents secured? Could they topple? Those paintings and mirrors in your bedroom: They're going to come crashing down. The broken glass is going to fly everywhere. At work, those drop ceilings are going to collapse. Computer monitors and anything on a shelf is going to go flying. There's a decent chance you're going to get hurt. Help is going to be a long time in coming. You don't have a go bag or an emergency first aid kit. Your chances of survival just decreased significantly. And they're about to get worse.

Probably at least one of your neighbors has a gas stove. Say the line ruptures and there's a fire. You've lost power and water. The streets around you have buckled, so no emergency crews can get there to help. You can't call them anyway, because the few remaining cell phone towers are so overwhelmed with calls that yours is blocked. That fire is going to rage and grow. Can you evacuate the area?

Those same jammed cell phone towers mean you can't text or call your kids, either. Where do they go to school? Is that building safe? Unless you live in California, there's a good chance it hasn't been built to seismic code and that the school hasn't practiced any earthquake drills with its students. There's going to be chaos. Maybe even fatalities. Say you are well enough to set out on foot. The downed power lines and toppled buildings are going to make the going exceedingly slow. If you have to cross a bridge or you live near a dam or a levee, you may be totally out of luck. But you're determined: You stick with it. What's happening in the meantime? Assuming your children are still alive and relatively unharmed, do they know what to do if you are separated and can't communicate with one another? If there is no adult to guide them, what decisions will they make, particularly if they are hurt and afraid? What if they wander away from the school? Even if you get there, you are going to have no way to find them. If they are there and trapped, what then? The police and first responders aren't going to be able to help—at least not until it may be too late.

This is a grim scenario for sure, but it's not a sensational one. As you've seen throughout this book, few if any regions of this country are prepared for a big quake. If it happens this year or the next or the one after that, there are going to be significant casualties. It's a roll of the dice, really, as to whether you and the people who mean the most to you will be some of them or not. Even if you survive, it could be a week before help gets to you. If you're like most Americans, you're not going to have what you need to get through that time.

But it doesn't have to be that way. Go back to the beginning of this scenario again and imagine it with a new paradigm in place.

It's 2:00 P.M. You're at work; your kids are still at school. Suddenly, all the cell phones around you start beeping: It's the federal earthquake early warning alert. You've got thirty seconds to take cover. You do. Across town, your children's school, which taxpayers agreed to renovate for seismic risk, receives the same alert. They've drilled for this. The kids get under their desks. A few minutes after the shaking stops, their teacher gives the all clear. The classrooms are strewn with some books and laptops, but everyone is okay. You still can't reach your kids by phone, but that's okay too: They know to make the short walk to the community center, where you will meet them as soon as you can. If they can't get there, they know to stay at the school. Maybe you've picked a particular spot there in the event that they can't stay inside and don't have a grown-up to help. The basketball court, say. The statue of the school mascot. You arrive with your go bag: Even if you all can't get home, you've got food and water and warm clothes—maybe even some Mad Libs to keep the little ones busy.

And, because your community has partnered with others in emergency management training, help is going to get to you pretty fast, especially since the bridges are passable, since the airport is up and running after a day of inspections. The local hospitals have been waiting for this event, so they have mobile triage centers in place within hours. If you or someone in your family is injured, help is within your reach. You have earthquake insurance, so even if your house is damaged or destroyed, it'll get fixed.

Is this quake still a disaster? Yes. And maybe not everyone will make it. But the scope of that disaster, on a regional level, is going to be one that you all can weather together. You'll bounce back. You are resilient.

We need to believe earthquake scientists when they tell us that the big one is coming. While it would be great if we find the technology to predict it, chances are that's not going to be possible in time. So it falls to all of us, both as individuals and as communities, to confront the risk this future event poses. If you live in Florida or North Dakota, that risk may be small enough to warrant little action (though certainly other natural disasters like tornados and hurricanes are more than real enough for you to take many of the same preventative steps you would if you lived in a more earthquake-prone zone). If you live on the Eastern Seaboard, if you live in a lot of the Midwest, or on the West Coast or in Utah or Oklahoma or Texas or Alaska or Hawaii, your risk is real. It's up to you to be ready for it.

ACKNOWLEDGMENTS

Works of popular science like this one are written entirely be-cause of the patience and good graces of scientists. I am deeply indebted to each and every one who appears in this book: the geologists and seismologists, physicists and engineers, who spent countless hours answering questions, sharing their research, and explaining concepts both simple and complex. They are too numer-ous to list individually here, but spare a very kind thought for each and every one of them as you come across their names in the pages of this book: They all took time away from the very important work that they do to make this project possible. I am particularly grateful for Emily Brodsky, Heather Savage, Todd Halihan, Brian Atwater, and David Yamaguchi, who extended hospitality and counsel far beyond what one could expect of even the most gener-ous scientist. I also want to thank the entire staff of Green Moun-tain College's Griswold Library, who filled dozens and dozens of interlibrary loan requests in record time: Without them, the research required for this project would have taken decades.

Equally important is the brilliant community of writers that surrounds me. Special thanks to Janine DeBaise and Camille Dungy, who both served as wingman for legs of this tour, along with Paul Bogard, Bill Roorbach, Murray Carpenter, Brian Kevin, Jennifer Sahn, Melissa Falcon, and Kate Christensen, who offered perspective and support throughout the process. I first wrote about fracking for an anthology called *Fracture* edited by Taylor Brorby and Stefanie Brook Trout; it's an important voice on this very contentious subject, and well worth the time to read and digest. My essay on earthquake acoustics appeared in *Ecotone,* one of the best serials out there, and was skillfully edited by Anna Lena Phillips Bell, who is as wise and kind as she is talented.

Stephanie Wade suggested the idea of approaching this book as a road trip, and I'm so glad that she did. Thanks also to all of my friends and family who provided love and wisdom in so many ways: Margot Carpenter of Hartdale Maps for the conference swag that began this project; Rene Fraze for the morning drive-time talks; Sarah and Drew Moore, who always asked about the project; Chris, Dawn, Hayden, and Josiah, who always had ideas and fun questions; and especially my parents, to whom this book is dedicated.

After a decade of working together, my agent, Wendy Strothman, seems as much family as any of them. She is equal parts fairy godmother and ruthless editor: the perfect combination in this business. Special thanks to her and Lauren MacLeod at the Strothman Agency for all that they do on behalf of their clients: I quite literally could not do this work without them. Thanks also to Stephen Morrow, Madeline Newquist, Mary Beth Constant (the world's finest—and most entertaining—copy editor), and the entire staff of Dutton for shepherding this project from nascent idea to the very lovely volume you now hold in your hands.

SOURCES

PROLOGUE

American Society of Civil Engineers. "2016 Report Card for America's Infrastructure," accessed November 23, 2016. http://www.infrastructurereportcard.org.

Federal Emergency Management Agency. *Earthquake Hazards Mitigation Handbook*. Washington, DC, 2002.

"The National Earthquake Hazards Reduction Program: Past, Present and Future." Hearing before the Subcommittee on Research, Committee on Science, House of Representatives, May 8, 2003. Serial no. 108-14.

US Geological Survey. "Earthquake Catalog," accessed February 19, 2017. https://earthquake.usgs.gov/earthquakes/search.

THEIR CAMPSITE, OUR CORE

Dzurisin, Daniel, Charles W. Wicks, and Michael P. Poland. "History of Surface Displacements at the Yellowstone Caldera, Wyoming, from Leveling Surveys and InSAR Observations." US Geological Survey professional paper 1788, v. 1.1, last modified October 10, 2012.

Ganchin, Y. V., Scott B. Smithson, I. B. Morozov, et al. "Seismic Studies around the Kola Superdeep Borehole, Russia." *Tectonophysics* 288, nos. 1–4 (30 March 1998): 1–16.

Hsu, Kenneth Jinghwa. *Challenger at Sea: A Ship that Revolutionized Earth Science*. Princeton, NJ: Princeton University Press, 1992.

Lowenstern, Jacob B., Robert B. Smith, and David P. Hill. "Monitoring Super-Volcanoes: Geophysical and Geochemical Signals at Yellowstone and Other Large Caldera Systems." *Philosophical Transactions of the Royal Society A* 364, no. 1845 (15 August 2006): 2055–2072.

Mastin, Larry G., Alexa R. Van Eaton, and Jacob B. Lowenstern. "Modeling Ash Fall Distribution from a Yellowstone Supereruption." *Geochemistry, Geophysics, Geosystems* 15, no. 8 (August 2014): 3459–3475.

Singh, Satish C., Nugroho Hananto, Yanfang Qin, et al. "The Discovery of a Conjugate System of Faults in the Wharton Basin Intraplate Deformation Zone." *Science Advances* 3, no. 1 (4 January 2017).

Smith, Robert, and Lee J. Siegel. *Windows into the Earth: The Geologic Story of Yellowstone and Grand Teton National Parks.* New York: Oxford University Press, 2000.

Stein, Seth, Carol Stein, Jonas Kley, et al. "New Insights into North America's Midcontinent Rift." *Eos* 97 (August 2016).

A BEAUTIFUL PLACE FOR AN EARTHQUAKE

Black, George. *Empire of Shadows: The Epic Story of Yellowstone.* New York: St. Martin's Press, 2012.

Bozeman Daily Chronicle, August 19, 1959.

Gilbert, G. K. *Salt Lake Daily Tribune*, September, 20, 1883.

Lewis, Meriwether, and William Clark. *The Journals of Lewis and Clark.* Edited by Bernard DeVoto. New York: Mariner Books, 1997.

Lyell, Charles. *Principles of Geology.* New York: Penguin, 1997.

McPhee, John. *Basin and Range.* New York: Farrar, Straus and Giroux, 1982.

Smith, Robert, and Lee J. Siegel. *Windows into the Earth: The Geologic Story of Yellowstone and Grand Teton National Parks.* New York: Oxford University Press, 2000.

Thon, Anita Painter. *Shaken in the Night: A Survivor's Story from the Yellowstone Earthquake of 1959.* CreateSpace Independent Publishing Platform, 2014.

US Geological Survey. "The Hebgen Lake, Montana, Earthquake of August 17, 1959." US Geological Survey professional paper 435, 1963.

Yellowstone Volcano Observatory. "Summary of 2010 Madison Plateau, Yellowstone Earthquake Swarm." https://volcanoes.usgs.gov/volcanoes/yellowstone/monitoring_swarm2010.html.

COFFEE IN SALT LAKE

Canham, Matt. "Salt Lake City Prison Sites Could Get Hit by Tsunami-like Wave." *Salt Lake Tribune*, December 17, 2014.

Coen, Deborah R. *The Earthquake Observers: Disaster Science from Lisbon to Richter.* Chicago: University of Chicago Press, 2013.

Dewey, James, and Perry Byerly. "The Early History of Seismometry (to 1900)." *Bulletin of the Seismological Society of America* 59, no. 1 (February 1969): 183–227.

Earthquake Engineering Research Institute. "Scenario for a Magnitude 7.0 Earthquake on the Wasatch Fault–Salt Lake City Segment." Federal Emergency Management Agency, 2015.

Machette, Michael N., Stephen F. Personius, and Alan R. Nelson. "The Wasatch Fault Zone, Utah—Segmentation and History of Holocene Earthquakes." *Journal of Structural Geology* 13, no. 2 (1991): 137–149.

Marsell, R. E. *The Wasatch Fault Zone in North Central Utah.* Utah Geological Society, 1964.

McPhee, John. *Annals of the Former World.* Farrar, Strauss, and Giroux, 2000.

Working Group on Utah Earthquake Probabilities (WGUEP). "Earthquake Probabilities for the Wasatch Front Region in Utah, Idaho, and Wyoming." Utah Geological Survey Miscellaneous Publication 16-3, 2016.

OUR FLOATING WORLD

Briggs, Peter. *200,000,000 Years beneath the Sea: The Story of the Glomar Challenger, the Ship That Unlocked the Secrets of the Oceans and Their Continents.* New York: Holt, Rinehart and Winston, 1971.

Frankel, Henry. *The Continental Drift Controversy: Wegener and the Early Debate,* reprint edition. Cambridge University Press, 2017.

Frisch, Wolfgang, Martin Meschede, and Ronald C. Blakey. *Plate Tectonics: Continental Drift and Mountain Building.* London: Springer, 2011.

Hazen, Robert M. *The Story of Earth: The First 4.5 Billion Years, from Stardust to Living Planet.* New York: Viking, 2012.

Lynch, David K. *Field Guide to the San Andreas Fault,* reprint edition. El Cajon, CA: Sunbelt Publications, 2015.

Oreskes, Naomi. *Plate Tectonics: An Insider's History of the Modern Theory of the Earth.* Boulder, CO: Westview Press, 2002.

Wegener, Alfred. *The Origin of Continents and Oceans.* Mineola, NY: Dover Publications, 2011.

COLD STORAGE

Almeida, R., J. Chester, F. Chester, T. J. Waller, and D. Kirschner. "Lithology and Structure of SAFOD Phase I Core Samples." *Earth and Planetary Science. Letters* 310 (2005): 131–144.

Dvorak, John. *Earthquake Storms: The Fascinating History and Volatile Future of the San Andreas Fault.* New York: Pegasus Books, 2014.

Fradkin, Philip L. *Magnitude 8: Earthquakes and Life along the San Andreas Fault.* Berkeley: University of California Press, 1999.

New York Times, August 11, 1884.

Oeser, Erhard. "Historical Earthquake Theories from Aristotle to Kant." In *Historical Earthquakes in Central Europe,* edited by Rudolf Gutdeutsch. Vienna: Geologische Bundesanstalt, 1992.

Richards, Paul G., and John Zavales. "Seismological Methods for Monitoring a CTBT: The Technical Issues Arising in Early Negotiations." In *Monitoring a Comprehensive Test Ban Treaty,* edited by Eystein S. Husebye and Anton M. Dainty. Dordrecht, Netherlands: Kluwer Academic Publishers, 1996.

US Congress, Office of Technology Assessment. *Seismic Verification of Nuclear Testing Treaties*, OTA-ISC-361. Washington, DC: US Government Printing Office, May 1988.

Winberry, J. Paul, Sridhar Anandakrishnan, Richard B. Alley, Robert A. Bindschadler, and Matt A. King. "Basal Mechanics of Ice Streams: Insights from the Stick-Slip Motion of Whillans Ice Stream, West Antarctica." *Journal of Geophysical Research* 114, no. 1 (March 2009). http://onlinelibrary.wiley .com/doi/10.1029/2008JF001035/full. Accessed March 11, 2017.

———. "Nucleation and Seismic Tremor Associated with the Glacial Earthquakes of Whillans Ice Stream, Antarctica." *Geophysical Research Letters* 2, no. 2 (January 2013): 312–315.

Zoback, M., S. Hickman, and W. Ellsworth. "Scientific Drilling into the San Andreas Fault Zone: An Overview of SAFOD's First Five Years." *Scientific Drilling* 11 (2011): 14–28.

DAM BUSTING WITH SENSURROUND G-FORCES!

Abd-El Monsef, Hesham, Scot E. Smith, and Kama Darwish. "Impacts of the Aswan High Dam after 50 Years." *Water Resources Management* 29, no. 6 (April 2015): 1873–1885.

Electric Power Research Institute (EPRI). *Central and Eastern United States Seismic Source Characterization for Nuclear Facilities: Review for Reservoir-Induced Seismicity (RIS) in the Southeast and Earthquakes in South Carolina near the 1886 Charleston Earthquake*. Palo Alto, CA, 2015.

Hadfield, Peter. "River of the Dammed." *New Scientist* 224 (November 2014): 46–49.

Hiltzik, Michael. *Colossus: The Turbulent, Thrilling Saga of the Building of Hoover Dam*. New York: Free Press, 2011.

Logan, Malcolm H. "Effect of the Earthquake of March 27, 1964, on the Eklutna Hydroelectric Project, Anchorage, Alaska." US Geological Survey professional paper 545-A, 1967.

Rastogi, B. K., and H. K. Gupta. *Dams and Earthquakes*. Elsevier Scientific Publishing Co., 1976.

Ridley, Matt. "First, the Bad News: We Can Cause Earthquakes." *Wall Street Journal*, November 19, 2011, C4.

Roger, Jedediah S. "Auburn Dam. Auburn Folsom Unit. American River Division Central Valley Project." Bureau of Reclamation Historic Reclamation Projects, 2009.

Rogers, A. M., and W. H. K. Lee. "Seismic Study of Earthquakes in the Lake Mead, Nevada-Arizona Region." *Bulletin of the Seismological Society of America* 66, no. 5 (October 1976): 1657–1681.

Theilen-Willige, Barbara, Sainath P. Aher, Praveen B. Gawali, and Laxmi B. Venkata. "Seismic Hazard Analysis along Koyna Dam Area, Western Maharashtra, India: A Contribution of Remote Sensing and GIS." *Geosciences* 6, no. 2 (2016): 20.

SOURCES

THE BIG MUDDY

Bradbury, John. *Travels in the Interior of America*. London: Sherwood, Neely, and Jones, 1819.

Dow, Lorenzo. *History of Cosmopolite, or the Four Volumes of Lorenzo Dow's Journal*, ed. Peggy Dow. Wheeling, VA: Joshua Martin, 1848.

Feldman, Jay. *When the Mississippi Ran Backwards: Empire, Intrigue, Murder, and the New Madrid Earthquakes*. New York: Free Press, 2005.

Jabr, Ferris. "Quake Escape." *New Scientist* 213, no. 2847 (January 2012): 34–37.

Marshall, Howard. *The New Madrid Earthquakes*. Columbia: University of Missouri Press, 1976.

Mueller, Karl, Susan E. Hough, and Roger Bilham. "Analysing the 1811–1812 New Madrid Earthquakes with Recent Instrumentally Recorded Aftershocks." *Nature* 429, no. 6989 (May 2004): 284–288.

Rusch, Elizabeth. "Earthquake Central." *Smithsonian* 42, no. 8 (December 2011): 18–21.

Tuttle, Martitia P. "Earth Science: New Madrid in Motion." *Nature* 435, no. 7045 (June 2005): 1037–1039.

Valencius, Conevery Bolton. *The Lost History of the New Madrid Earthquakes*. Chicago: University of Chicago Press, 2015.

THE LUCKY FRIDAY MINE

Gibowicz, Slawomir Jerzy, and Andrzej Kijko. *An Introduction to Mining Seismology*. San Diego: Academic Press, 1994.

Klose, Christian D. "Geomechanical Modeling of the Nucleation Process of Australia's 1989 M5.6 Newcastle Earthquake." *Earth and Planetary Science Letters* 256, nos. 3–4 (April 2007): 547–553.

Mark, Christopher. "Coal Bursts in the Deep Longwall Mines of the United States." *International Journal of Coal Science & Technology* 3, no. 1 (March 2016): 1–9.

McGarr, A., David Simpson, and L. Seeber. "Case Histories in Induced and Triggered Seismicity." In *International Handbook of Earthquake and Engineering Seismology*, edited by William H. K. Lee, Hiroo Kanamori, Paul C. Jennings, and Carl Kisslinger, 647–661. London: Academic Press, 2002.

Swanson, Peter, Brian Kenner, and Todd Krahenbuhl. "Seismic Event Data Acquisition and Processing: Distribution and Coordination across PC-Based Networks." Paper presented at the Application of Computers and Operations Research in the Mineral Industry: Proceedings of the 30th International Symposium, Phoenix, AZ, February 25–27, 2002.

Swanson, Peter L., and C. D. Sines. "Characteristics of Mining-Induced Seismicity and Rock Bursting in a Deep Hard-Rock Mine." US Department of the Interior, Bureau of Mines, Report of Investigations 9393, 1991.

Wang, Jimin, Xionghui Zeng, and Jifang Zhou. "Practices on Rockburst Prevention and Control in Headrace Tunnels of Jinping II Hydropower Station." *Journal of Rock Mechanics and Geotechnical Engineering* 4, no. 3 (25 September 2012), 258–268.

Whyatt J. K., W. Blake, T. J. Williams, and B. G. White. "60 Years of Rockburst-
ing in the Coeur d'Alene District of Northern Idaho, USA: Lessons Learned
and Remaining Issues." Paper presented at the 109th Annual Exhibit and
Meeting of the Society for Mining, Metallurgy, and Exploration, Phoenix,
AZ, February 25–27, 2002.

Whyatt, J. K., F. M. Jenkins, and M. K. Larson. "In Situ Stress at the Lucky Friday Mine (in Four Parts)." US Department of the Interior, Bureau of Mines,
Report of Investigations 9582, 1995.

DIGGING BACK EAST

Costain, John K., and G. A. Bollinger. "Review: Research Results in Hydroseismicity from 1987 to 2009." *Bulletin of the Seismological Society of America*
100, no. 5A (October 2010): 1841–1858.

Ruhl, Arthur. "Building New York's Subway." *Century Magazine*, October
1902, 894–907.

Sayles, R. W. "Earthquakes and Rainfall." *Bulletin of the Seismological Society
of America* 3, no. 2 (June 1913): 51–56.

Yoders, Jeffrey. "Guidelines for Development around Tunnels and Seismic Com-
pliance of Underground Structures." Bentley Systems, April 2012.

OUR INGENIOUS WELLS

Coen, Deborah R. *The Earthquake Observers: Disaster Science from Lisbon to
Richter*. Chicago: University of Chicago Press, 2013.

Fleeger, Gary M., and Daniel J. Goode. "Hydrologic Effects of the Pymatuning Earthquake of September 25, 1998, in Northwestern Pennsylvania." US Geological
Survey, Water-Resources Investigations Report 99-4170, August 1999.

Gorokhovich, Y. "Abandonment of Minoan Palaces on Crete in Relation to the
Earthquake Induced Changes in Groundwater Supply." *Journal of Archaeo-
logical Science* 32 (2005): 217–222.

Gorokhovich, Y., and G. Fleeger. "Pymatuning Earthquake in Pennsylvania and
Late Minoan Crisis on Crete." *Water Science and Technology: Water Supply*
7, no. 1 (March 2007): 245–251.

Hough, Susan E., and Morgan Page. "Potentially Induced Earthquakes during
the Early Twentieth Century in the Los Angeles Basin." *Bulletin of the Seismological Society of America* 106, no. 6 (December 2016): 2419–2435.

Independent Record (Helena, MT), July 2, 1963.

US Geological Survey. *2015 Guide to Earthquakes from Fracking, Hydraulic
Fracturing, and Shale Gas—Underground Wastewater Disposal*. Washington,
DC: Department of the Interior, 2015.

THE EARTHQUAKE FIGHTERS OF OKLAHOMA

Evans, David M. "The Denver Area Earthquakes and the Rocky Mountain Arsenal Disposal Well." *Mountain Geologist* 3, no.1 (1966): 23–36.

Galchen, Rivka. "Weather Underground." *The New Yorker* (April 13, 2015): 34–38.

Keranen, Katie M., Heather M. Savage, Geoffrey A. Abers, and Elizabeth S. Cochran. "Potentially Induced Earthquakes in Oklahoma, USA: Links between Wastewater Injection and the 2011 M 5.7 Earthquake Sequence." *Geology* 41, no. 6 (June 2013): 699–703.

Preliminary Report on the Northstar 1 Class II Injection Well and the Seismic Events in the Youngstown, Ohio, Area. Ohio Department of Natural Resources, March 2012.

Rall Walsh, F., and Mark D. Zoback. "Probabilistic Assessment of Potential Fault Slip Related to Injection Induced Earthquakes: Application to North-Central Oklahoma, USA." *Geology* 44, no. 12 (December 2016): 991–994.

Witze, Alexandra. "Artificial Quakes Shake Oklahoma: Earthquakes Linked to Oil and Gas Preparations Prompt Further Research into Human-Induced Seismic Hazards." *Nature* 520 (April 2015): 418–420.

TANK FARMS ON THE PRAIRIE

Bachmann, C. E., W. Foxall, and T. Daley. "Comparing Induced Seismicity on Different Scales." Paper presented at the Proceedings: 39th Workshop on Geothermal Reservoir Engineering Stanford University, Stanford, California, February 24–26, 2014.

Cappa, Frédéric, and Jonny Rutqvist. "Seismic Rupture and Ground Accelerations Induced by CO_2 Injection in the Shallow Crust." *Geophysical Journal International* 190, no. 3 (September 2012): 1784–1789.

Johnston, Jill E., Emily Werder, Daniel Sebastian, et al. "Wastewater Disposal Wells, Fracking, and Environmental Injustice in Southern Texas." *American Journal of Public Health* 106, no. 3 (March 2016), 550–556.

Lubick, Naomi. "California Heats Up over Natural Steam." *Scientific American*, December 10, 2001.

Petersen, Mark D., Charles S. Mueller, Morgan P. Moschetti, et al. "Incorporating Induced Seismicity in the 2014 United States National Seismic Hazard Model—Results of 2014 Workshop and Sensitivity Studies." US Geological Survey Open-File Report 2015–1070.

Stein, Ross. "Earthquake Conversations." *Scientific American*, January 2003.

Thomson, Jason. "Oklahoma Earthquake: Are Oil Storage Terminals Safe from Shaking?" *The Christian Science Monitor*, November 7, 2016, accessed March 17, 2017. http://www.csmonitor.com/USA/2016/1107/Oklahoma-earthquake-Are -oil-storage-terminals-safe-from-shaking.

US Environmental Protection Agency. *Assessment of the Potential Impacts of Hydraulic Fracturing for Oil and Gas on Drinking Water Resources.* EPA/600/R-15/047a, June 2015.

COOL STUFF TO STORE IN MINES

Alley, William M., and Rosemarie Alley. *Too Hot to Touch: The Problem of High-Level Nuclear Waste.* New York: Cambridge University Press, 2013.

D'Agata, John. *About a Mountain.* New York: W. W. Norton, 2010.

Hoffman, David. *The Dead Hand: The Untold Story of the Cold War Arms Race and Its Dangerous Legacy.* New York: Anchor, 2010.

Malley, Marjorie. *Radioactivity: A History of a Mysterious Science.* New York: Oxford University Press, 2011.

McFarlane, Allison, and Rodney Ewing. *Uncertainty Underground: Yucca Mountain and the Nation's High-Level Nuclear Waste.* Cambridge, MA: MIT Press, 2006.

Schlosser, Eric. *Command and Control: Nuclear Weapons, the Damascus Accident, and the Illusion of Safety.* New York: Penguin, 2014.

Smith, Ken, John G. Anderson, and Amy J. Smiecinski. "Seismicity in the Vicinity of Yucca Mountain, Nevada, for the Period October 1, 2004 to September 30, 2006." Digital Scholarship@UNLV, November 2007. http://digitalscholarship .unlv.edu/yucca_mtn_pubs/58.

Stuckless, John, and Robert A. Levich. "The Road to Yucca Mountain—Evolution of Nuclear Waste Disposal in the United States." *Environmental & Engineering Geoscience Journal* 22, no. 1 (February 2016): 1–25.

US Congress, Office of Technology Assessment. *Seismic Verification of Nuclear Testing Treaties*, OTA-ISC-361. Washington, DC: US Government Printing Office, May 1988.

Vartabedian, Ralph. "Nuclear Accident in New Mexico Ranks among the Costliest in U.S. History." *Los Angeles Times*, August 22, 2016.

Wynn, Jeffrey C., and Eugene H. Roseboom. "Role of Geophysics in Identifying and Characterizing Sites for High-Level Nuclear Waste Repositories." *Journal of Geophysical Research* 92, no. B8 (July 1987): 7787–7796.

BLOCKS AND BLOCKS

Baker, W. S. *Portraits of Washington.* Philadelphia: Robert M. Lindsay, 1887.

Cheng, M. H., M. D. Kohler, and T. H. Heaton. "Prediction of Wave Propagation in Buildings Using Data from a Single Seismometer." *Bulletin of the Seismological Society of America* 105 (2015): 107–119.

FEMA. *Promoting the Adoption and Enforcement of Seismic Building Codes: A Guidebook for State Earthquake and Mitigation Managers.* CreateSpace Independent Publishing Platform, 2013.

Gilbert, Daniel, and Sandi Doughton. "Buildings That Kill: The Earthquake Danger Lawmakers Have Ignored for Decades." *Seattle Times*, May 14, 2016.

Hansen, Gladys, and Emmet Condon. *Denial of Disaster: The Untold Story and Photographs of the 1906 San Francisco Earthquake.* Petaluma, CA: Cameron and Co., 1989.

International Code Council. *2015 International Building Code.* ICC (Cengage), 2014.

Kueht, Erin, and Mary Beth D. Hueste. "Impact of Code Requirements in the Central United States: Seismic Performance Assessment of a Reinforced Concrete Building." *Journal of Structural Engineering* 135, no. 4 (April 2009): 404–413.

Lindeburg, Michael R., and Kurt M. McMullin. *Seismic Design of Building Structures.* Belmont, CA: Professional Publications, 2014.

McIntosh, R. D., and S. Pezeshk. "Comparison of Recent U.S. Seismic Codes." *Journal of Structural Engineering* 123, no. 8 (August 1997): 993–1001.

GIMME SHELTER

Coen, Deborah R. *The Earthquake Observers: Disaster Science from Lisbon to Richter*. Chicago: University of Chicago Press, 2014.

Harada, Nahako, Jun Shigemura, Masaaki Tanichi, et al. "Mental Health and Psychological Impacts from the 2011 Great East Japan Earthquake Disaster: A Systematic Literature Review." *Disaster and Military Medicine* 1, no. 17 (2 September 2015).

Ripley, Amanda. *The Unthinkable: Who Survives When Disaster Strikes—and Why*. New York: Harmony, 2009.

Solnit, Rebecca. *A Paradise Built in Hell: The Extraordinary Communities That Arise in Disaster*. New York: Penguin, 2010.

Tsujiuchi, Takuya. "Mental Health Impact of the Fukushima Nuclear Disaster: Post-Traumatic Stress and Psycho-Socio-Economic Factors." Fukushima Global Communication Programme Working Paper Series, no. 8, December 2015.

Williams, Susan Millar, and Stephen G. Hoffius. *Upheaval in Charleston: Earthquake and Murder on the Eve of Jim Crow*. Athens: University of Georgia Press, 2012.

THE SCHOOL

Atwater, Brian, Musumi-Rokkaku Satoko, Satake Kenji, et al. *The Orphan Tsunami of 1700: Japanese Clues to a Parent Earthquake in North America*, 2nd edition. Seattle: University of Washington Press, 2016.

Doughton, Sandi. *Full-Rip 9.0: The Next Big Earthquake in the Pacific Northwest*, reprint edition. Seattle: Sasquatch Books, 2014.

Henderson, Bonnie. *The Next Tsunami: Living on a Restless Coast*. Corvallis: Oregon State University Press, 2014.

Molesky, Mark. *This Gulf of Fire: The Great Lisbon Earthquake, or Apocalypse in the Age of Science and Reason*, reprint edition. New York: Vintage, 2016.

Thomson, Jerry. *Cascadia's Fault: The Coming Earthquake and Tsunami That Could Devastate North America*, reprint edition. Berkeley, CA: Counterpoint, 2012.

Winchester, Simon. *Krakatoa: The Day the World Exploded: August 27, 1883*. New York: Harper Perennial, 2005.

Wood, Nathan J., Jeanne Jones, Seth Spielman, and Mathew C. Schmidtlein. "Community Clusters of Tsunami Vulnerability in the US Pacific Northwest." *PNAS* 112, no. 17 (28 April 2015): 5354–5359.

PREDICTING THE UNPREDICTABLE

Bakun, W. H., J. Bredehoeft, R. O. Burford, et al. "Parkfield Earthquake Prediction Scenarios and Response Plans." US Geological Survey Open-File Report 86-365, June 12, 1986.

Brodsky, Emily E., and Thorne Lay. "Recognizing Foreshocks from the 1 April 2014 Chile Earthquake." *Science* 344, no. 6185 (16 May 2014): 700–702.

Geller, Robert J. "Earthquake Prediction: A Critical Review." *Geophysical Journal International* 131, no. 3 (December 1997): 425–450.

Hough, Susan. *Predicting the Unpredictable: The Tumultuous Science of Earthquake Prediction*, reprint edition. Princeton, NJ: Princeton University Press, 2016.

Johnson, Brian Fisher. "Earthquake Prediction: Gone and Back Again." *Earth*, April 7, 2009.

Lindh, Allan G. "Earthquake Prediction Comes of Age." *MIT Technology Review*, February/March 1990, 44–51.

Michael, Andrew, Paul Reasenberg, Peter H. Stauffer, and James W. Hendley II. "Quake Forecasting—An Emerging Capability." US Geological Survey Fact Sheet 242-95, March 1995.

Milne, John. *Earthquakes and Other Earth Movements*. Palala Press, 2016.

Spall, Henry, and David W. Simpson, eds. *The Soviet-American Exchange in Earthquake Prediction*. US Geological Survey Open-File Report 81-1150, 1981.

IS THE SKY FALLING?

Dollar, John. "The Man Who Predicted an Earthquake." *Guardian*, April 5, 2010.

Gasparini, Paolo, Gaetano Manfredi, and Jochen Zschau (eds.). *Earthquake Early Warning Systems*. New York: Springer, 2007.

Jordon, Thomas H. "Don't Blame Italian Seismologists for Quake Deaths." *New Scientist*, September 21, 2011.

Stucchi, Massimiliano, Rui Pinho, and Massimo Cocco. "After the L'Aquila Trial." *Seismological Research Letters* 87, no. 3 (May 2016): 591–596.

Tributsch, Helmut. *When the Snakes Awake*. Cambridge, MA: MIT Press, 1982.

Wyss, Max, and Silvia Peppoloni (eds.). *Geoethics: Ethical Challenges and Case Studies in Earth Sciences*. Waltham, MA: Elsevier, 2014.

YOUR TWO-MINUTE WARNING

Cheng, M., S. Wu, and T. Heaton. "Earthquake Early Warning Application to Buildings." *Engineering Structures* 60 (2014): 155–164.

Earle, Paul S., Daniel C. Bowden, and Michelle Guy. "Twitter Earthquake Detection: Earthquake Monitoring in a Social World." *Annals of Geophysics* 54, no. 6 (2011).

Hough, Susan E. "A Seismological Retrospective of the Brady-Spence Prediction." *Seismological Research Letters* 81, no. 1 (January/February 2010).

Wenzel, Friedemann, and Jochen Zschau (eds.). *Early Warning for Geological Disasters: Scientific Methods and Current Practice*. New York: Springer, 2013.

INDEX

ABOUT THE AUTHOR

Kathryn Miles is an acclaimed journalist and writer-in-residence for Green Mountain College, as well as a faculty member for Chatham University's MFA program. With a BA in philosophy from Saint Louis University and a PhD in English from the University of Delaware, Kate is also a scholar-in-residence for the Maine Humanities Council and a member of the Terrain.org editorial board. Her work has appeared in *The Best American Essays*, *Popular Mechanics*, *Outside*, and *The New York Times*.